Chemistry and Nutritional Effects of Capsicum

Food Chemistry, Function and Analysis

Series editors:
Gary Williamson, *Monash University, Australia*
Alejandro G. Marangoni, *University of Guelph, Canada*
Graham A. Bonwick, *AgriFoodX Limited, UK*
Catherine S. Birch, *AgriFoodX Limited, UK*

Titles in the series:

1: Food Biosensors
2: Sensing Techniques for Food Safety and Quality Control
3: Edible Oil Structuring: Concepts, Methods and Applications
4: Food Irradiation Technologies: Concepts, Applications and Outcomes
5: Non-extractable Polyphenols and Carotenoids: Importance in Human Nutrition and Health
6: Cereal Grain-based Functional Foods: Carbohydrate and Phytochemical Components
7: Steviol Glycosides: Cultivation, Processing, Analysis and Applications in Food
8: Legumes: Nutritional Quality, Processing and Potential Health Benefits
9: Tomato Chemistry, Industrial Processing and Product Development
10: Food Contact Materials Analysis: Mass Spectrometry Techniques
11: Vitamin E: Chemistry and Nutritional Benefits
12: Anthocyanins from Natural Sources: Exploiting Targeted Delivery for Improved Health
13: Carotenoid Esters in Foods: Physical, Chemical and Biological Properties
14: Eggs as Functional Foods and Nutraceuticals for Human Health
15: Rapid Antibody-based Technologies in Food Analysis
16: DNA Techniques to Verify Food Authenticity: Applications in Food Fraud
17: Advanced Gas Chromatography in Food Analysis
18: Handbook of Food Structure Development
19: Mitigating Contamination from Food Processing
20: Biogenic Amines in Food: Analysis, Occurrence and Toxicity
21: Nutrition and Cancer Prevention: From Molecular Mechanisms to Dietary Recommendations
22: Health Claims and Food Labelling
23: Nutraceuticals and Human Health: The Food-to-supplement Paradigm
24: Nutritional Signalling Pathway Activities in Obesity and Diabetes
25: The Chemistry and Bioactive Components of Turmeric
26: Foodomics
27: Food Proteins and Peptides: Emerging Biofunctions, Food and Biomaterial Applications
28: Handbook of Antioxidant Methodology: Approaches to Activity Determination
29: Fats and Associated Compounds: Consumption and Human Health

30: Oral Processing and Consumer Perception: Biophysics, Food Microstructures and Health
31: Development of Trans-free Lipid Systems and their Use in Food Products
32: Advanced Spectroscopic Techniques for Food Quality
33: Berries and Berry Bioactive Compounds in Promoting Health
34: Metabolism of Nutrients by Gut Microbiota
35: Bio- and Nano-sensing Technologies for Food Processing and Packaging
36: Edible Fungi: Chemical Composition, Nutrition and Health Effects
37: Chemistry and Nutritional Effects of Capsicum

How to obtain future titles on publication:
A standing order plan is available for this series. A standing order will bring delivery of each new volume immediately on publication.

For further information please contact:
Book Sales Department, Royal Society of Chemistry, Thomas Graham House, Science Park, Milton Road, Cambridge, CB4 0WF, UK
Telephone: +44 (0)1223 420066, Fax: +44 (0)1223 420247.
Email: booksales@rsc.org
Visit our website at www.rsc.org/books

Chemistry and Nutritional Effects of Capsicum

By

Valdir Florencio da Veiga, Jr
Military Institute of Engineering, Brazil
Email: valdir.veiga@gmail.com

Larissa Silveira Moreira Wiedemann
Federal University of Amazonas, Brazil
Email: larissasm@yahoo.com.br

Claudio Pereira de Araujo, Jr
Military Institute of Engineering, Brazil
Email: cpajr50@gmail.com

and

Ananda da Silva Antonio
Federal University of Amazonas, Brazil
Federal University of Rio de Janeiro, Brazil
Email: ananda.antonio@gmail.com

ROYAL SOCIETY OF **CHEMISTRY**

Food Chemistry, Function and Analysis No. 37

Print ISBN: 978-1-78801-750-3
PDF ISBN: 978-1-83916-064-6
EPUB ISBN: 978-1-83916-065-3
Print ISSN: 2398-0656
Electronic ISSN: 2398-0664

A catalogue record for this book is available from the British Library

The Royal Society of Chemistry is a charity, registered in England and Wales, Number 207890, and a company incorporated in England by Royal Charter (Registered No. RC000524), registered office: Burlington House, Piccadilly, London W1J 0BA, UK, Telephone: +44 (0) 20 7437 8656.

Visit our website at www.rsc.org/books

Printed in the United Kingdom by CPI Group (UK) Ltd, Croydon, CR0 4YY, UK

Foreword

The future need for the (bio)chemical and allied industries and their outputs is undeniable because they are directly connected to impending global megatrends in population rise, urbanisation, industrialisation, construction, health care and sanitation, food security and scarcity, and resource depletion.

However, we cannot continue to proceed with the (bio)chemical and manufacturing processes of the past, which rely on crude oil as a feedstock and are both resource-intensive and resource-depleting, and problematic at the end of their life cycle.

We need to adopt green and sustainable processes that are commensurate with circular rather than linear economies, do "more with less" and abide with the UN Sustainable Development Goals to achieve a sustainable twenty-first century. We need to explore the diversity of renewable feedstocks for their wealth of (bio)chemical applications. There is no shortage of renewable feedstocks, but simply a matter of selecting the right ones.

This book eloquently explores chilli peppers as an excellent example of a renewable feedstock rich in chemical complexity and natural products beneficial for health, food, cosmetics and other non-lethal applications. The book is written by eminent experts and will serve as a brilliant educational resource for both experts and non-experts on the wonders of chilli peppers. It brings together chemistry and biology and is an enabler for a sustainable twenty-first century.

Avtar S. Matharu
Professor of Green Chemistry
Deputy Director, Green Chemistry Centre of Excellence
University of York, York, UK
avtar.matharu@york.ac.uk

Food Chemistry, Function and Analysis No. 37
Chemistry and Nutritional Effects of Capsicum
By Valdir Florencio da Veiga, Jr, Larissa Silveira Moreira Wiedemann, Claudio Pereira de Araujo, Jr and Ananda da Silva Antonio

Published by the Royal Society of Chemistry, www.rsc.org

Preface

Chilli peppers are one of the world's most famous spices. Almost every adult has eaten a *Capsicum* pepper, usually without knowing its scientific name. *Capsicum* peppers are much more than just a spicy taste – they are also a formidable natural product. This book aims to help you understand from a scientific perspective how and why *Capsicum* peppers became so popular and are so integrated with human history.

This book provides the chemical and biological background to the intrinsic characteristics of peppers, such as colour, smell and taste. The chemical composition of peppers is reviewed in detail and used to explain their biological properties. The popularization of peppers in early history was a result of their remarkable taste and medicinal properties. *Capsicum* peppers are a unique natural product containing an exclusive class of chemicals: the capsaicinoids. The compounds in this class of chemicals are responsible for the unique spicy taste of peppers.

This book discusses the chemical and biological features of *Capsicum* that have enabled the development of several different technologies, including cosmetics and, more controversially, non-lethal weapons. We believe that this book will help to build a solid bank of knowledge for researchers developing natural products based on the diversity of *Capsicum*.

Valdir Florencio da Veiga, Jr
Larissa Silveira Moreira Wiedemann
Claudio Pereira de Araujo, Jr
Ananda da Silva Antonio

Food Chemistry, Function and Analysis No. 37
Chemistry and Nutritional Effects of Capsicum
By Valdir Florencio da Veiga, Jr, Larissa Silveira Moreira Wiedemann,
Claudio Pereira de Araujo, Jr and Ananda da Silva Antonio

Published by the Royal Society of Chemistry, www.rsc.org

Contents

Food Chemistry, Function and Analysis No. 37
Chemistry and Nutritional Effects of Capsicum
By Valdir Florencio da Veiga, Jr, Larissa Silveira Moreira Wiedemann, Claudio Pereira de Araujo, Jr and Ananda da Silva Antonio

Published by the Royal Society of Chemistry, www.rsc.org

CHAPTER 1

Origin and Evolution of Capsicum

1.1 Introduction

What are chilli peppers? When we think about this term, most of us picture a small red fruit with a conical shape, no longer than our fingers, with a burning taste (Figure 1.1). Most people also imagine chilli peppers simply as a cooking ingredient. People also misidentify black pepper as a chilli pepper because the term "chilli pepper" is often wrongly used to denote any hot or spicy pepper. The American English term "chili" comes from the name "chilli", a Nahuatl word from the Aztec civilisation, which was used specifically for hot *Capsicum* peppers. This name became more popular as *Capsicum* spread around the world and gained several translations. The term "chilli pepper" is one of more than 200 common names given to the hot species of the genus *Capsicum*.[1]

Capsicum encompasses so much more than "chilli peppers". The genus contains a huge biodiversity within its nearly 40 species and they have many more applications than just cooking spicy foods. It is estimated that *Capsicum* made an appearance in human history around 7500 BC and started to be cultivated between 5200 and 3400 BC.[2] *Capsicum* was not only used for its taste, but also had other ethnobotanical usages. Records dating back to the sixth century report the use of *Capsicum* fruits in medicinal preparations. Chilli peppers are used to treat wounds, relieve pain (*e.g.* stomach aches, headaches and indigestion), reduce fever, as an anti-infective agent and to treat hypertension when chewed. Chilli pepper smoke was used as a punishment for children and as a weapon in hunting and warfare. The smoke of chilli peppers causes the same burning sensation

Food Chemistry, Function and Analysis No. 37
Chemistry and Nutritional Effects of Capsicum
By Valdir Florencio da Veiga, Jr, Larissa Silveira Moreira Wiedemann, Claudio Pereira de Araujo, Jr and Ananda da Silva Antonio

Published by the Royal Society of Chemistry, www.rsc.org

Figure 1.1 The conical fruits known as chilli peppers. Image Credit: Carlos Magno da Silva Antonio.

in the eyes, lungs, throat, nose and skin as occurs in the mouth when eaten.[3,4] Can you imagine how painful this could it be? Nowadays, the use of chilli peppers as a weapon is restricted, as discussed in Chapter 10.

In addition to these ancient uses of *Capsicum*, modern society still uses peppers as a source of bioactive compounds for drug development and as food additives. Which brings us to the question: how did *Capsicum* become such a remarkable source of bioproducts worldwide? It received attention because of its flavour and its use in treating ailments. However, the Age of Discovery and the expeditions of Christopher Columbus were defining points in the spread of chilli peppers across the world and their worldwide commercialisation. In the pre-Columbian era, *Capsicum* was dispersed by both birds (which also like the taste of chilli peppers) and humans, including the Inca and Aztec civilisations.[1] Despite the pre-Columbian exchange of chilli peppers by different civilisations, there is no consensus about how *Capsicum* was introduced to Asia, although archaeological records confirm the origin of *Capsicum* in the Americas.[4,5]

Origin is an important issue in everything with which we interact. For instance, a known origin can aid in understanding how a plant interacts and evolves in the environment and enables us to develop the best technology to exploit and preserve it. We would therefore like to take you on a trip back to the origin of chilli peppers.

1.2 *Capsicum*, Chili or Chilli?

The terminology used to refer to the *Capsicum* species includes more than 200 names, including common names, botanical terminology and variety identification. Some of these (*e.g.* the terms "chilli peppers", "ají", "paprika", "pepper" and "chili") are more common and are used synonymously to define any pungent or powdered fruit. Nowadays, two terms are popularly used to distinguish pungent from non-pungent *Capsicum* species, which are, respectively, chilli peppers and sweet bell peppers (Figure 1.2).

The name *Capsicum* was first suggested as a botanical term to define the genus in 1543. The term was derived from the Latin word "capsa" and the Greek term "kapto", which mean "box" and "to bite". Both terms are believed to have been chosen because they allude to the shape of the fruit and the pungency of a bite. Between 1543 and 1753, several attempts were made to taxonomically describe *Capsicum* species, as highlighted in Morrison's *Plantarum Historiae Universalis Oxoniensis* (1699), which described 33 different peppers, and Tourneforte's description in 1700, which is considered to be responsible for the consolidation of the term "*Capsicum*" to define the genus. Linnaeus (1753 and 1767) and Diószegi and Fazekas (1807) gave the first concise taxonomic descriptions of the five domesticated species of *Capsicum*.[6,7]

The other terms used to describe *Capsicum* species are derived from their local popular names. For instance, the term "ají" comes from the word "axi" in an extinct Arawak dialect, whereas the term "chilli" is originally from the Nahuartl Aztec dialect, which has been translated into several languages, gaining the variations "chili" and "chile". The removal of one "l" was an adaptation of the word for American English, while the additional "e" is

Figure 1.2 *Capsicum* peppers with a bell shape: the sweet bell peppers. Image Credit: Bárbara Elisabeth Teixeira Costa.

related to the Hispanic spelling. Some varieties, such as chiltepin (*C. annuum* var. *glabriusculum*), also have popular names derived from extinct dialects. Chiltepin comes from the words "chil" and "tepin" of the Nahuartl Aztec dialect. In addition, the term "pepper", which follows "chili", is derived from the Sanskrit "pippali", the same name given to *Piper longum* (black pepper). The use of "pepper" for either *Capsicum* and *Piper* is based on their similar taste properties: the burning feeling.

1.3 The Age of Discovery: Routes for the Dissemination of Peppers

Humans have always sought different ways to season their food and the Indian region was the origin of the spices most widely used today. Some of these spices were known by the Greeks, whereas others were only discovered by Europeans at the time of the Crusades (1096–1291).[8,9] We start our journey in the Middle Ages and then travel through one of the most beautiful times in human history: the Renaissance and the discovery of the New World.

After the end of the Crusades, several of the supply chains that had been paved to transport and trade goods to fighting armies enabled commerce between Europe and the Middle East, including faraway countries such as India and China. Crossing the Middle East to reach Europe, the Spice Route (Figure 1.3) brought precious products to the northern countries. With the fall of Constantinople on 29 May 1453, the whole world changed. Several waves of migration of writers, poets, lecturers, musicians and artists left the city to move west. This marked the end of the Middle Ages and led to the Renaissance. Architects, geographers, academics and astronomers brought the finest maps from Africa and Asia to Europe and also technologies from

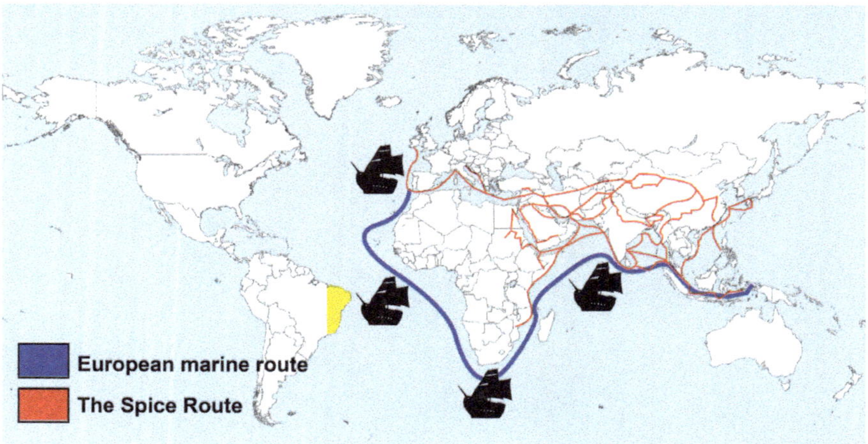

Figure 1.3 Map of the Spice Route and the European marine route for spices. Image Credit: Carlos Magno da Silva Antonio.

Ancient Greece and the Orient. This brought a new wave of knowledge in the second half of the fifteenth century. As Arabs started controlling the Spice Route, Venice, the centre of European commerce, saw prices soaring and Indian products became much more expensive.[10,11] This included the spices sold in macerated, dried and powdered forms, which were indispensable in European cuisine. An alternative route to India was devised: across the ocean.

Portugal, facing out to the Atlantic Ocean, became a Kingdom in 1139. At the beginning of the fourteenth century, the nautical sciences in Iberia were superior to any other region of the globe. Portugal began to improve its sailing ships (caravels) and conquered several Atlantic islands, such as the archipelagos of Madeira in 1419 and the Azores in 1427, and the west coast of Africa from 1434, including the Cabo Verde archipelago in 1460.

Launching their ships in pursuit of new sea trade routes was a natural path for the Portuguese and Spanish kingdoms. They had conquered several overseas colonies and were enriched by commerce, achieved through their enhanced navigational knowledge of the Atlantic Ocean and new maps from Africa obtained from Middle Eastern astronomers and geographers. This resulted in the discovery of the Americas by Christopher Columbus in 1492, the maritime route to India bypassing Africa by Vasco da Gama in 1498, and the first circumnavigation of the world by Fernão de Magalhães and Juan Sebastián Elcano in 1522. The new routes to India and the discovery of the Americas and many Indonesian islands enabled the establishment of a strong and exclusive commercial path to the Iberian countries, generating long-lasting empires in Asia and the Americas (Figure 1.3).

In addition to the known Indian, Chinese and Indonesian peppers and cinnamons, many other new products were brought to Europe and spread throughout the world. From the Americas, the world was introduced to maize, avocado, potato, cotton, rubber, chicle, coca, tobacco, peanuts, guarana, *açaí* fruit, pineapple, blueberry, guava, chilli peppers and other *Capsicum* species. Portuguese and Spanish navigators also found the known Indian black pepper (*P. nigrum*) in the tropical Americas, as well as other peppers from the genus *Capsicum* that were unique to the New World.[9,12,13] These species were sent to Europe by the first explorers and were introduced to Spain by Christopher Columbus in 1493. In the same year, an entry from Christopher Columbus' diaries from his last day in Haiti states: "There is also a lot of axi, which is their pepper, of a kind as precious as pepper, and none of the natives eat anything without it, because they find it very sound. Fifty caravels could be loaded with it each year in Haiti".

Diego Álvarez Chanca, a doctor onboard the Columbus expedition, shipped the first *Capsicum* plants to Spain in 1494 with a letter describing their medicinal effects. Chanca's letter also included the following statement: "They use, to season it, a vegetable called aji, which they also employ to give a sharp taste to the fish and such birds as they can catch, of the infinite variety there are in this island, dishes of which they prepare in different ways".

Peppers from *Capsicum* genus were taken by the Portuguese from South America to Europe, Africa and Asia and, in 1542, they were cultivated in India, where they were promptly adopted for consumption. In the sixteenth century, a variety of pepper grown in Goa, India was called pernambuco (named after a state in northeastern Brazil), documenting its origin and intensifying the relationship between European populations and indigenous peoples. The cultivation of this pepper was later recorded in China in 1700.[8,9,14,15]

In the period between 1492 and 1600, shipping routes allowed pepper species to travel the world, especially from Brazil to Portugal, Africa and, later, Asia. There was a broad distribution of the genus *Capsicum* from the Mediterranean region to England in 1548 and to Central Europe, where it was immediately accepted. After its introduction to Europe, the genus was dispersed to various parts of the world, including North America. This new condiment was incorporated almost instantaneously into the cuisine of countries in Europe, Africa and Asia. Even in the East, despite being the land of spices, where it might have been difficult to accept a new spicy substance, *Capsicum* became widespread. It was called "poor pepper" because it was incomparably cheaper than black pepper and had a similar burning, spicy taste. Five centuries after the discovery of the Americas, peppers came to dominate the spice trade in both tropical and temperate regions.

1.4 Archaeological Records and Historical Usage

Latin America can be considered as the centre of origin for species of the genus *Capsicum*.[16] Long before the arrival of Columbus in the Americas, *Capsicum* were widely used in Central and South America, the Caribbean and Mexico, with at least 27 reported species. The origin and history of the *Capsicum* genus is complex due to its worldwide dispersion by humans. The origin of the genus is usually studied over the five species that are the most cultivated today: *Capsicum annuum*, *Capsicum baccatum*, *Capsicum frutescens*, *Capsicum chinense* and *Capsicum pubescens*. Several archaeological records indicate that these peppers were consumed in the Andean regions of Peru and in Mexico from at least 8600–5600 BC.[17]

Archaeological finds are either illustrations created by ancient civilisations, remains of fruit parts in human artefacts or preserved plant parts. In the Codex Mendoza, which contains records of Aztec civilisations, there are several illustrations of their tribute system that include pictures of chilli peppers. The Aztec civilisation is known to have used chilli peppers to spice up a cocoa bean drink known as tchocoatl, a precursor of chocolate, and pottery and embroidery adorned with chilli peppers dating back to 500 AD has been found in Peru.

The first proof of the presence of peppers in the human diet, estimated to be from 7500 BC, was found in Tehuacán, Mexico (Figure 1.4). The discovery was in Coxcatlán cave, an archaeological site considered to be key to our understanding of the early history of plant domestication. Several remains of

Figure 1.4 Archaeological sites with the most ancient *Capsicum* fossils. Image Credit: Carlos Magno da Silva Antonio.

C. annuum fruits were discovered in the cave, along with maize and agave. *C. annuum* is currently the most commercialised *Capsicum* species, with its centre of diversity thought to be Central America and Mexico.

Another finding from the transition from gathering to a cultivation-based economy was made in Guitarrero cave, Peru. The cave is another famous archaeological site with evidence of humans dating back to about 8000 BC. The *Capsicum* spp. fossil found in Guitarrero cave was a whole fruit with its calyx estimated to date back to nearly 6500 BC. The fossil had a layer just after its calyx that indicated it was a fully domesticated variety.

Peru has several archaeological sites from this transitioning economy. Sites such as Huaca Pietra have the oldest records of plant domestication, including one of the oldest fossils of maize in South America. Huaca Pietra contained seeds of *Capsicum*, mainly of *C. baccatum*, dating from 9300 to 3800 BC. The evidence from archaeological sites in Peru indicate that *Capsicum* were domesticated by humans as early as 10 000 BC (Figure 1.4).[8,18] The Huaca Pietra site and others along the Andean coast are related to the Inca empire. According to historians, chilli peppers were of more value than money in commerce and were considered to be holy plants in Inca creation myths. There is a tale in Inca mythology known as Brother Chile Pepper, which retells the origin and ascension of the Inca empire through the story of four very strong brothers. The brother representing chilli peppers was Agar-Uchu and the association with the spicy fruit that gives taste to many of the dishes in their cuisine represents the joy and contentment in life after the creation of the Incan empire. Several varieties of cultivated chilli pepper were described by the major historian of the Inca empire, Garcilaso de la Vega, including: *rocot uchu*, the broad pepper, a *C. pubescens* variety; *kellu uchu*, the lemon drop, a *C. baccatum* variety; and *chinchi uchu*, a cherry-like pod fruit of *C. chinense*.

Huaca Pietra and the nearby Paredones site have archaeological evidence of *C. baccatum*, *C. pubescens*, *C. chinense* and *C. frutescens* dating back to 7600–6500 BC. However, the majority of samples were of *C. baccatum* dating to 4500 BC, representing the ability of the local civilisation to domesticate this species. Still in Peru, archaeological finds in Punta Grande and Ancón identified seeds and fruits of *C. baccatum* with morphological traits of domestication dating back to 2000–500 BC.[19,20] *C. baccatum* was largely cultivated in the Andean coastal region. This species, referred to as "ají" in this region, is one of the most important ingredients in Bolivian and Peruvian cuisine. In Brazil, three varieties of these species are known: pepper-to-pepper (*C. baccatum*), coumari (*C. baccatum* var. *praetermissum*) and cambuci (*C. baccatum* var. *pendulum*).[21] Despite its popularity in its centre of dispersion, *C. baccatum* was one of the least commercialised peppers during the Age of Discovery.

Fossils of *C. frutescens* fruits were found in Cahuachi, a site used by the Nazca people (southern Peru) during peregrination. Curiously, it was still possible to detect capsaicin, a pungent metabolite of *Capsicum*, in the fossils from this site. Pollen microfossils of *C. frutescens* were found in sediments from a drained agricultural field in La Tigra, Venezuela. The pollen samples date to 1000–450 BC and, because the field covered >35 ha, indicate the agricultural exploitation of chilli peppers.

The most ancient archaeological finds of *C. chinense* date to about 8600 BC from the Guitarrero cave.[20] Fossil remains of a domesticated *C. chinense* fruit were also found in the Casma Valley, Peru, along with remains of maize, beans and potatoes. Evidence of wild varieties of *C. chinense* were found in Roraima, Brazil, reinforcing the idea that *C. chinense* arrived in the Andean coastal region after domestication in Brazil. Despite the fact that most people believe the name *C. chinense* was given to suggest a possible Asian origin, this is not true. The archaeological finds corroborate that *C. chinense* is of Latin America origin and, more precisely, an Amazon basin species. Several of the famous *C. chinense* cultivars (*e.g.* habanero and biquinho) are native to Amazonia. The indigenous people of the Amazon basin had plenty of uses for this species and are currently considered to be responsible for its domestication.[16,22]

Capsicum chinense is known among chilli peppers as the species that contains the hottest varieties. In 2010, an internet challenge emerged in which people shared videos of themselves biting a *C. chinense* variety known as bhut jolokia, one of the hottest peppers in the world. This variety has been cultivated in India for the last 7000 years. Such hot cultivars also have medicinal properties and can be used to heal wounds, treat intestinal disorders, relieve muscle pain and toothache, and control arthritic pain; they are used by indigenous communities in the Amazon basin.[4,23,24]

Other records that reconstruct the domestication of *Capsicum* in the pre-Columbian era were found in the buried remains of Cerén, El Salvador. This was a farming community that had a similar fate to Pompeii and was buried by volcanic ash. Remains of *C. annuum* seeds and pedicels, dating to about 1400 BC, were found at this site. Their locations within the site suggest that the fruits were being stored for further use. More recent evidence for the popularity of *C. annuum* was found in the archaeological site at En Bas Saline, Haiti, a Chican Ostionoid settlement. Seed and pedicel remains dating to 600 BC were found in positions within the settlement that indicate their possible use in trading and cultivation in home gardens.

As an ancient botanical species, chilli peppers have passed through human history with many applications. Archaeological finds of *Capsicum* are not only of fruit, flower and seed fossils, but also handmade artefacts, such as food receptacles and vessels.[3] In Brazil, by the time of the arrival of Europeans, the cultivation of peppers was the most common practice among indigenous peoples. In fact, Central and South America were the birthplace of domesticated peppers. Reports of peppers cultivated by Brazilian indigenous peoples in the sixteenth century were made by the German Hans Staden. Reports also describe trade between indigenous peoples and French explorers, whose ships landed on the Brazilian coast in search of local products. Peppers were highly prized plants among the indigenous peoples, together with maize and cassava.

Expeditions during the sixteenth century in Brazil have several narratives regarding the importance of chilli peppers for the population at the time. Quotations include descriptions such as:[25] "Peppers in great quantity…" (Sylveira 1624) and peppers would be "…as indispensably necessary to the natives as salt for the whites" (Humboldt 1841). Among the first and foremost scientific records of peppers is the work of Leonhartus Fuchsius from 1543, named *Stirring History*. It presents a very detailed description and three drawings of a pepper tree.

In Europe, *Capsicum* arrived by the Spice Route, but not as a spice. The first samples that arrived in Europe were considered rare and expensive items and were taken by botanists and aristocrats. European nobles in the sixteenth century primarily used *Capsicum* spp. as ornamental plants in their gardens. As it started being cultivated in the countryside of Europe by farmers who could not afford Asian black pepper, it was no long considered interesting as an ornamental plant and instead was known as "the poor people's peppers". Only in the eighteenth century were *Capsicum* peppers fully consolidated in Europe as a spice in several traditional recipes.[26]

Capsicum spp. has many applications in our society, from ornamental plants to medicine. There are also some very peculiar ethnobotanic uses of this genus involving superstitions. Brazilian indigenous communities use chilli peppers to keep bad spirits away when entering forests. Indigenous people gift a new friend with a pepper to make the friendship last, but it cannot be delivered by hand and must be placed in a location where the new friend will find it. In Europe and Mexico, chilli peppers are also used as a symbol of virility and as an aphrodisiac to "spice up" romantic relationships.[25,26]

1.5 Geographical Distribution

The geographical distribution of the genus *Capsicum* has four major epicentres where the species originated naturally, all of them within the American continent (Figure 1.5).[9] Nowadays, *Capsicum* is cultivated throughout the world.

The centre of origin of *Capsicum* is thought to be in the Andes region, with a clockwise expansion in South America to northeastern Brazil, southeastern Brazil, back to western South America and entering Central America with the most recent clade (Figure 1.5).[9,13,27] As *Capsicum* expanded geographically,

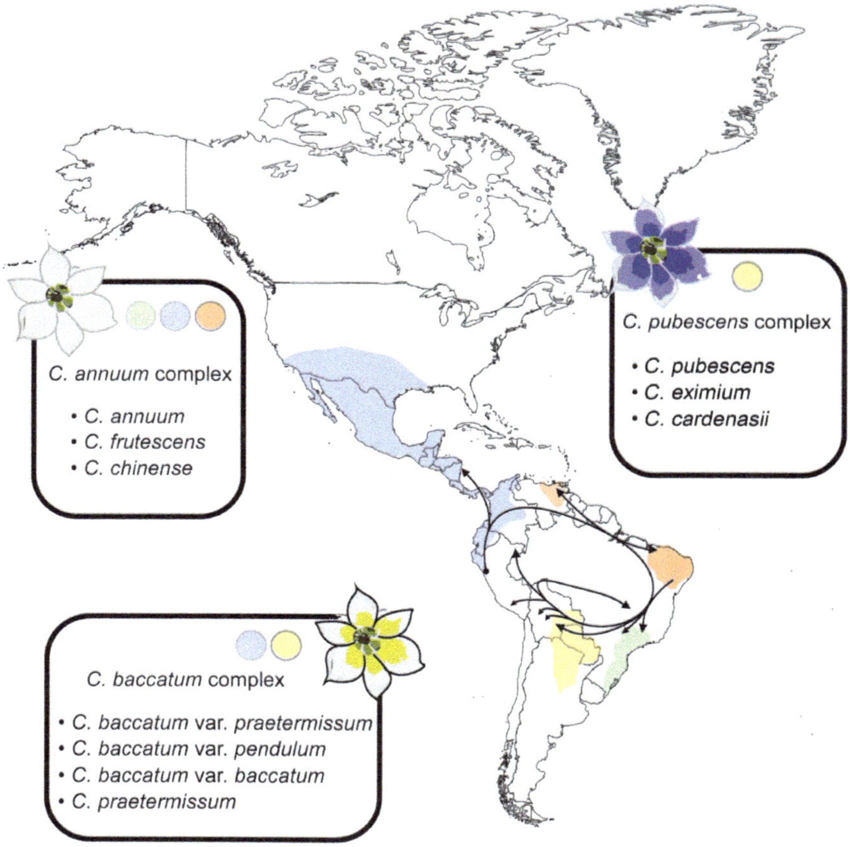

Figure 1.5 Natural epicentres of occurrence of *Capsicum*. Image Credit: Carlos Magno da Silva Antonio.

the distinct climatic conditions in each location promoted the development of species complexes – that is, groups of different species with very similar morphological features that often make it difficult to distinguish between them by eye. Three complexes were formed through gene crossing, which, coincidentally, can be morphologically characterised by the colour pattern of their flowers.[27,28] These complexes are known as the *C. annuum* complex, the *C. baccatum* complex and the *C. pubescens* complex.

The first complex, known as the *C. annuum* complex, possibly originated in the lower tropical regions of South and Central America due to its affinity for wetter environments. This complex includes species such as *C. annuum* var. *annuum*, *C. annuum* var. *aviculare*, *C. annuum* var. *glabriusculum*, *Capsicum chacoense*, *Capsicum galapagoense*, *C. frutescens* and *C. chinense*. Morphologically, these species are similar with dull white flowers.[27,29,30] There is some dispute as to whether *C. chacoense* is part of the *C. annuum* or the *C. baccatum* complex.[29,30]

Interesting varieties of *C. annuum* include jalapeño, poblano and ancho. In addition, cayenne and paprika spices are derived from this species. Jalapeño peppers are used in the preparation of a less spicy variety of Tabasco sauce. This *C. annuum* species has cultivated forms that appear to have reached South America, possibly after the Spanish conquests.[31] The centre of diversity of the cultivated form of *C. annuum* var. *annuum* includes Asia and Latin America.[32] Wild populations of *C. annuum* are found from the southern USA and Mexico to Central America and northern South America.[6] *C. annuum* var. *glabriusculum* is found in the southern USA, Mexico, the Antilles, Belize, Honduras, El Salvador, Panama, Costa Rica, Guatemala, Suriname, Venezuela, Colombia, Ecuador, Peru, and the north and northeast of Brazil. This species is popularly known as the American bird pepper, turkey pepper, cayenne pepper, chile-tepin and chiltepin.

Although the *C. annuum* complex has its epicentre in South America, its species each have their own centre of origin. For instance, the species *C. annuum* is thought to be originally from Mexico (a little bit far from the epicentre of the complex, but only because it is the youngest species).[33] By contrast, the centre of origin of *C. frutescens* is thought to be in Peru. *C. frutescens* wild cultivars are widely distributed in the tropical and subtropical Americas, whereas domesticated cultivars are grown in the USA, Mexico, South and Central America, Africa, India, China and Japan. The tabasco cultivar is the sole representative of this species commonly grown outside the tropics.[32,34]

Capsicum chinense is the most widespread species of the *C. annuum* complex in tropical America. Its centre of origin is the Amazon basin, which currently holds its centre of biodiversity.[20] Some varieties of this species grown in Africa are reported to be the most pungent of all peppers.[34] *C. chinense* is the most commonly grown pepper in the West Indies, northern South America and the Amazon basin. In addition, some of its domesticated cultivars are grown in the USA, Mexico, Central America, Ecuador, Peru, Bolivia, Argentina, China and Japan.[22] The limited global popularity of the *C. chinense* species, compared with other *C. annuum* complex species, is most likely due to its discovery in South America after *C. annuum* had already been established on the European continent.[35]

The *C. baccatum* complex has characteristics related to dry environments and therefore may have originated in the south-central regions of Bolivia and Peru and, based on archaeological evidence, was probably domesticated in Peru around 2500 BC. It is commonly consumed in South America,[36] but is little known outside this region.[34] Some wild species of *C. baccatum* can be found in Mexico, India and Hawaii.[35] This complex mainly consists of white-flowered species with yellow dots or lines on the petals. Among the species, *C. baccatum* var. *pendulum* is cultivated in the USA, Mexico, Costa Rica, Colombia, Ecuador, Peru, Brazil, Bolivia, Paraguay, Chile, Argentina and India, whereas *C. baccatum* var. *baccatum* is cultivated in Colombia, Peru, Bolivia, Paraguay, south and southeast Brazil, and northern Argentina. The varieties *pendulum* and *baccatum* are popularly known as dedo de moça and cumari, respectively. *C. baccatum* is mainly consumed and commercialised in Peru, Bolivia and Brazil.

The *C. pubescens* complex, sometimes called the *C. emixia* complex, originated in Bolivia and extended across northern Central America and Mexico.[9] This complex consists of species such as *C. pubescens*, *Capsicum eximium* and *Capsicum cardenasii*, characterised by purple flowers.

The cultivation of *C. pubescens* (popular name rocoto), the only domesticated species in the complex, is centred in South America, Guatemala and southern Mexico at mid-elevations (1500–3000 m). Its most remarkable characteristic is that it requires a cold climate and can resist temperatures as low as -5 °C.

1.6 Concluding Remarks

The genus *Capsicum* is one of the oldest and most cultivated crops worldwide. Known by many names, it was also one of the major spices of the Spice Route and the Inca empire. Long before the Common Era, brought about by the exploration of marine routes, *Capsicum* peppers had become domesticated in their centre of diversity. Peru and Mexico have many archaeological sites in which the economic and cultural importance of the genus is clear. Despite the great importance of the Age of Discovery in the worldwide spread of *Capsicum*, it was already a common product before that time. The commercial routes travelled by foot from the centre of origin of the species are still unclear and the extensive commercialisation of the species makes it difficult to trace the ancestor species with accuracy. The human influence in the dispersion of *Capsicum* is so important that its existence within the major products of the Spice Route may have defined the world economy we know today. As an essential ingredient in traditional dishes and part of the human diet, *Capsicum* has played a part in motivating marine expeditions. Its taste and medicinal properties are both remarkable and unique, as you will see as you read this book.

References

1. N. Madala and M. K. Nutakki, *Int. J. Curr. Microbiol. Appl. Sci.*, 2020, **9**, 2532–2538.
2. F. A. Khan, T. Mahmood, M. Ali, A. Saeed and A. Maalik, *Nat. Prod. Res.*, 2014, **28**, 1267–1274.
3. T. G. Powis, E. G. Murrieta, R. Lesure, R. L. Bravo, L. Grivetti, H. Kucera and N. W. Gaikwad, *PLoS One*, 2013, **8**, e79013.
4. M. K. Meghvansi, S. Siddiqui, M. H. Khan, V. K. Gupta, M. G. Vairale, H. K. Gogoi and L. Singh, *J. Ethnopharmacol.*, 2010, **132**, 1–14.
5. S. Knapp, *Science*, 2007, **315**, 946–947.
6. D. A. Zhigila, A. A. Abdulrahaman, O. S. Kolawole and F. A. Oladele, *J. Bot.*, 2014, **2014**, 1–6.
7. G. Csilléry, *Acta Agron. Hung.*, 2006, **54**, 151–166.
8. L. Perry, R. Dickau, S. Zarrillo, I. Holst, D. M. Pearsall, D. R. Piperno, M. J. Berman, R. G. Cooke, K. Rademaker, A. J. Ranere, J. S. Raymond, D. H. Sandweiss, F. Scaramelli, K. Tarble and J. A. Zeidler, *Science*, 2007, **315**, 986–988.
9. E. A. Moscone, M. A. Scaldaferro, M. Grabiele, N. M. Cecchini, Y. S. García, R. Jarret, J. R. Daviña, D. A. Ducasse, G. E. Barboza and F. Ehrendorfer, *Acta Hortic.*, 2007, **745**, 137–169.
10. V. M. Russo, *Peppers: botany, production and uses*, CABI International 2012, Wallingford, 2012.
11. *Handbook of Eating and Drinking*, ed. H. L. Meiselman, Springer, Rockport, 2020.
12. V. S. Govindarajan, *Crit. Rev. Food Sci. Nutr.*, 1985, 22, 109–176.
13. C. Carrizo García, M. H. J. Barfuss, E. M. Sehr, G. E. Barboza, R. Samuel, E. A. Moscone and F. Ehrendorfer, *Ann. Bot.*, 2016, **118**, 35–51.
14. P. G. Smith, J. Charles and B. Heiser, *Bull. Torrey Bot. Club*, 1957, **84**, 413–420.
15. H. Padilha and R. Barbieri, *Aust. J. Basic Appl. Sci.*, 2016, **10**, 148–154.
16. F. J. B. Reifschneider, *Capsicum:pimentas e pimentões no Brasil*, Embrapa Comunicações para Transferência de Tecnologia, Brasília, 2000.
17. F. Nuez, M. J. Diez, J. J. Ruiz, P. F. Cordova, J. Costa, M. S. Catalá, A. González and J. A. Rodríguez, *Catálogo de semillias de pimiento*, Ministerio de Agricultura, Pesca y Alimentación, Instituto Nacional de Investigacíon y Tecnología Agraria y Alimentaria, Madri, 1998.
18. M. J. McLeod, S. I. Guttman and W. H. Eshbaugh, *Econ. Bot.*, 1982, **36**, 361–368.
19. V. W. D. Casali and F. A. A. Couto, *Origem e botânica de Capsicum*, Embrapa Hortaliças, Brasília, 1984.
20. C. R. Clement, M. de Cristo-Araújo, G. C. D'Eeckenbrugge, A. A. Pereira and D. Picanço-Rodrigues, *Diversity*, 2010, **2**, 72–106.
21. W. H. Eshbaugh, *Brittonia*, 1970, **22**, 31–43.
22. B. Pickersgill, *Evolution*, 1971, **25**, 683–691.
23. B. Singh, *Botanical Leads for Drug Discovery*, Springer, Singapore, 2020.

24. A. L. C. Roman, L. C. Ming, I. De Carvalho and M. D. G. P. Sablayrolles, *Bol. Mus. Para. Emilio Goeldi Ciencias Humanas*, 2011, **6**, 543–557.
25. H. R. Do Nascimento Filho, R. I. Barbosa and F. J. D. F. Luz, *Acta Amazonica*, 2007, **37**, 561–565.
26. E. Katz, in *Food, Imaginaries and Cultural Frontiers. Essays in honour of Helen Macbeth*, ed. F. X. Medina, Universidad de Guadalajara, Guadalajara, 2009, pp. 213–232.
27. B. M. Walsh and S. B. Hoot, *Int. J. Plant Sci.*, 2012, **162**, 1409–1418.
28. F. N. Viñals, R. G. Ortega and J. C. Garcia, *El cultivo de pimientos, chiles y ajies*, Mundi-Prensa, Madri, 1996.
29. C. K. Khoury, D. Carver, D. W. Barchenger, G. E. Barboza, M. van Zonneveld, R. Jarret, L. Bohs, M. Kantar, M. Uchanski, K. Mercer, G. P. Nabhan, P. W. Bosland and S. L. Greene, *Diversity Distrib.*, 2020, **26**, 209–225.
30. K. Shiragaki, S. Yokoi and T. Tezuka, *Horticulturae*, 2020, **6**, 1–13.
31. B. Pickersgill, in *Chromosome engineering in plants: genetics, breeding, evolution, part B*, ed. T. Tsuchiya and P. K. Gupta, Elsevier, Amsterdam, 1991, pp. 139–160.
32. D. W. Barchenger, P. Naresh and S. Kumar, in *The Capsicum Genome*, ed. N. Ramchiary and C. Kole, Springer, Cham, 2019, pp. 9–23.
33. F. D. Martin, J. Santiago and A. Cook, *Agric. Res.*, 1979, **16**, 200–218.
34. C. B. Heiser, in *Evolution of Crop Plants*, ed. N. W. Simmonds, Longman, London, 2nd edn, 1976, pp. 265–268.
35. W. H. Eshbaugh, in *New Crops*, ed. J. Janick and J. E. Simon, John Wiley and Sons, Inc., New York, 1993, pp. 132–139.
36. P. W. Bosland, *Capsicums: Innovative Uses of an Ancient Crop*, VA: ASHS Press, Arlington, 1996.

CHAPTER 2

Biological Features of Capsicum

2.1 Introduction

The *Capsicum* genus is one of the most important genera in the Solanaceae family and consists of 40 accepted species.[1] Native to South and Central America, but broadly commercialised and domesticated, the *Capsicum* genus is currently cultivated worldwide and contains hundreds of cultivars and varieties. Among its 40 species, only five are considered to be domesticated: *Capsicum annuum* L., *Capsicum chinense* Jacq., *Capsicum frutescens* L., *Capsicum baccatum* L. and *Capsicum pubescens* Ruiz & Pav.

We see several types of sweet and chilli peppers on a daily basis, but do we know how to distinguish between them? Although there are several key features, with so many shapes, colours, flavours and plant structures, it is not a simple task to identify each one of them (Figure 2.1).

The domestication of *Capsicum* species and the development of cultivars increased the complexity of genus systematics. *Capsicum* is currently a well-resolved genus, but it is hard to distinguish most of its species by morphology alone. Furthermore, the phylogenetic relationships have scarcely been elucidated as a result of the genetic diversity within this genus. These problems also make it difficult to establish new species.[2]

Domestication and breeding are essential in the economic trade of this genus, which is currently one of the most important commercial spices worldwide. Despite the issues associated with each systematics approach, this research cannot be stopped because we need to use this genus in a sustainable manner, conserving both wild and domesticated species. The sustainable use and preservation of species requires a clear understanding of the features that distinguish them: morphological, genetic and chemical. Knowing the diversity within each of these factors enables us to improve the

Food Chemistry, Function and Analysis No. 37
Chemistry and Nutritional Effects of Capsicum
By Valdir Florencio da Veiga, Jr, Larissa Silveira Moreira Wiedemann,
Claudio Pereira de Araujo, Jr and Ananda da Silva Antonio

Published by the Royal Society of Chemistry, www.rsc.org

Figure 2.1 The high variability of the morphological characteristics of *Capsicum* fruits. Image credit: Carlos Magno da Silva Antonio.

genetics of species by breeding cultivars adapted to a certain environment or resistant to a specific disease or pest.

2.2 Taxonomic Description

The name *Capsicum* comes from a Greek derivative of a Latin word meaning "to bite". This was supposed to reference the unique sensation experienced on biting a chilli pepper. At that time, the non-pungent species were poorly exploited. *Capsicum* peppers are known for their diversity of colour, shape, aroma and flavour, which also means that they have a notable morphological variety. *Capsicum* species can be classified based on their level of domestication.

- Domesticated: species selected by humans with a specific life cycle, morphological features and desired farming pattern. They are usually unable to survive in natural conditions. Examples include *C. annuum* L., *C. chinense* Jacq., *C. frutescens* L., *C. baccatum* L. and *C. pubescens* Ruiz & Pav.
- Semi-domesticated: species that have already experienced a certain level of genetic modification by human selection, but are in transition from a condition in which they can easily survive under natural conditions to full dependence on human farming. Examples include *Capsicum eximium* Hunz. and *Capsicum campylopodium* Sendtn.
- Wild: species that only occur in natural environments are unable to be farmed by humans. Examples include *Capsicum chacoense* Hunz., *Capsicum galapagoense* Hunz., *Capsicum praetermissum* Heiser & P. G. Sm. and *Capsicum tovarii* Eshbaugh. P. G. Sm. & Nickrent.

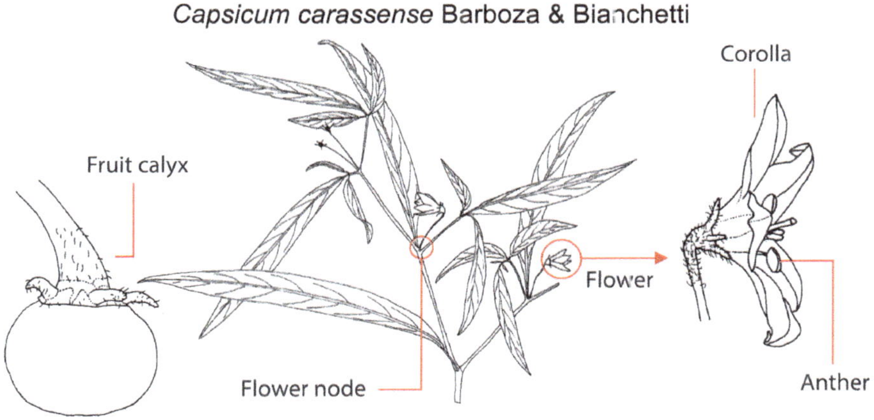

Figure 2.2 Representation of the major morphological features used in the identification of domesticated *Capsicum* species. Adapted from ref. 3, https://doi.org/10.3897/phytokeys.140.47071, under the terms of the CC BY 4.0 license, https://creativecommons.org/licenses/by/4.0/. Original drawn by L. Bianchetti.

Despite the great morphological variety within the *Capsicum* genus, the domesticated species can be easily identified and distinguished from each another. The major morphological identification markers are related to the characteristics of the petals (corolla), calyx and flower nodes (Figure 2.2).[3]

Corollas vary in colour and shape; the calyx can be long or small, smooth or contorted; and the flower may vary from one to seven per node (Table 2.1).[4,5] These morphological features are more effective in distinguishing the domesticated species of *Capsicum*. In general, the corolla and the colour and number of flowers per node are the minimal features required to distinguish the domesticated species.

Capsicum species occur as perennial shrubs, biennial herbaceous plants and arborescent trees. Flowers can present corollas as either stellate, rotate or campanulate. The colour of the corolla has been used to classify species into two groups (white and purple), but this classification cannot properly place the yellow-flowered species.

A morphological element usually applied in the species distinction of *Capsicum* fossils is the seed shape. A technical report produced by the Archaeobotany Laboratory of the University of California Berkeley evaluated 44 samples of domesticated *Capsicum* to identify seeds found in archaeological sites in Peru. Their results indicated that each of the five domesticated species has distinct crushed seed formats and beak prominences (a protrusion in the seed shape) (Figure 2.3). The other seed features identified in the same technical note to distinguish these *Capsicum* species were the texture of the seed coat, the beak angle, the ratio of the maximum seed length to the perpendicular width, the seed sphericity, the ratio of the

Table 2.1 Major morphological features used to differentiate domesticated *Capsicum* species.

Morphological feature	*Capsicum pubescens*	*Capsicum baccatum*	*Capsicum annuum*	*Capsicum chinense*	*Capsicum frutescens*
Corolla	Violet with white centre	Shades of white to yellow	Pure white to bluish white	Dull white	Greenish white
Seeds	Dark brown to black	Cream to yellow or black	Pale yellowish	Wavy, with colours varying from cream to yellow	Cream to yellow
Anthers	Purple to violet	Blue to purple	Blue to purple	Blue to violet	Blue to violet
Calyx	Conspicuous deltoid teeth nearly 1 mm long	Teeth present or absent; if present, teeth are >0.5 mm	Calyx teeth present or absent; if present, teeth are up to 0.5 mm	Calyx with no teeth	Calyx presenting few to no teeth
Flower	One or two per node	Cream with gold or green marks	One per node	Two or more per node	Two or more per node

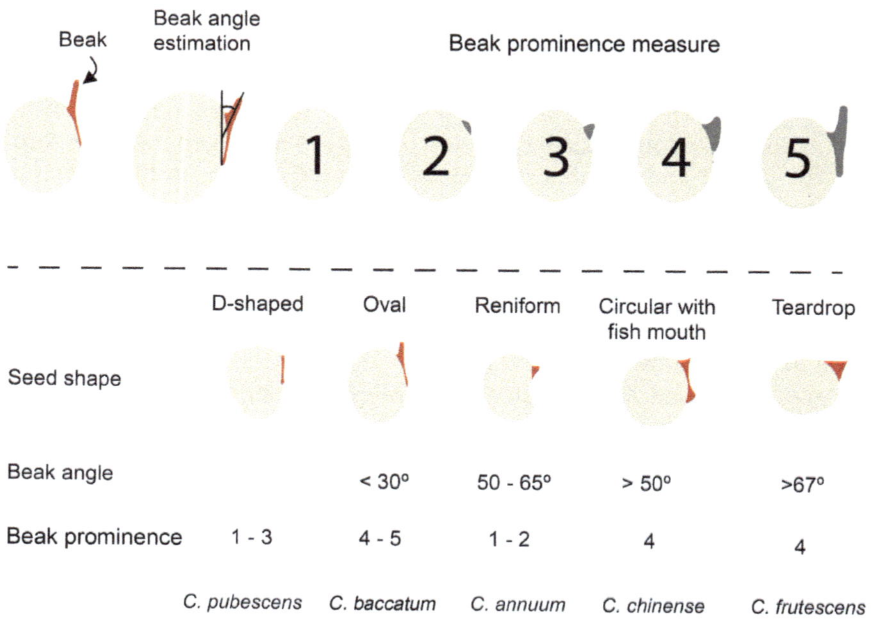

Figure 2.3 Crushed seed features used to distinguish *Capsicum* species. Image credit: Carlos Magno da Silva Antonio.

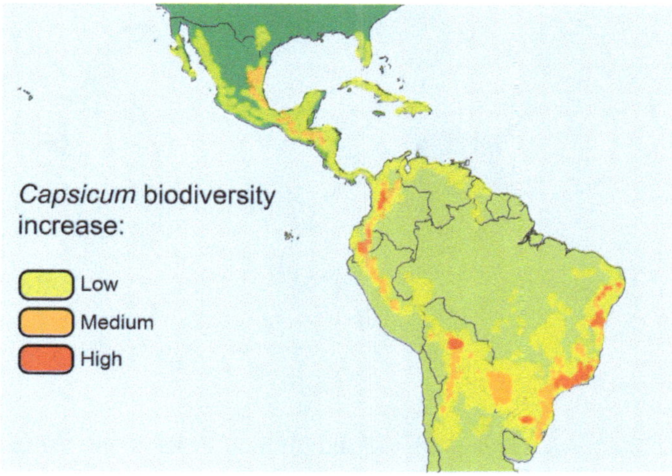

Figure 2.4 Regions with major occurrence of wild *Capsicum* species. Adapted from ref. 7, https://doi.org/10.1111/ddi.13008, under the terms of the CC BY 4.0 license, https://creativecommons.org/licenses/by/4.0/.

thickest portion of the seed testa to the thinnest portion and the sphericity of the attachment scar.[6]

The taxonomic identification of wild species is much more difficult. New wild species of *Capsicum* are still being discovered and many others gain new identifier keys to reduce ambiguous identification. The major biodiversity centres of wild *Capsicum* species are within the Andes and in southern Brazil, in the Atlantic forest (Figure 2.4).[7,8] Several identification keys for Andean and Atlantic forest wild species have been evaluated by Barboza and coworkers.[8,9]

2.2.1 *Capsicum annuum* var. *annuum* L.

Capsicum annuum is the type species of the genus, meaning it was the species first used to describe the characteristics of the *Capsicum* genus, mainly based on the anatomical profile of the flower and fruit (Figure 2.5). The characterisation of *C. annuum* as a type species does not mean that it is the most ancient species in the genus.

The *C. annuum* species is an annual herb or sub-shrub with several branches, varying from 0.5 to 1.5 m tall. The different cultivars and varieties of *C. annuum* can be classified by the characteristics of their fruit. Flowers grow individually, one per node, with the pedicel erect, pendulous or inclined and up to 3 cm long. The corolla is campanulate, usually white, without spots, and with a bluish anther. The corolla is 8–15 mm in diameter, containing five to seven stamens, an ovary 2(-4)-locular and a style filiform. In the fruits, the pedicels are up to 8 cm long with a cup-shaped calyx and five conspicuous teeth. The fruits are usually conical, but their colour and shape vary greatly within genotypes as a result of human selection for these distinct characteristics.[4,10,11]

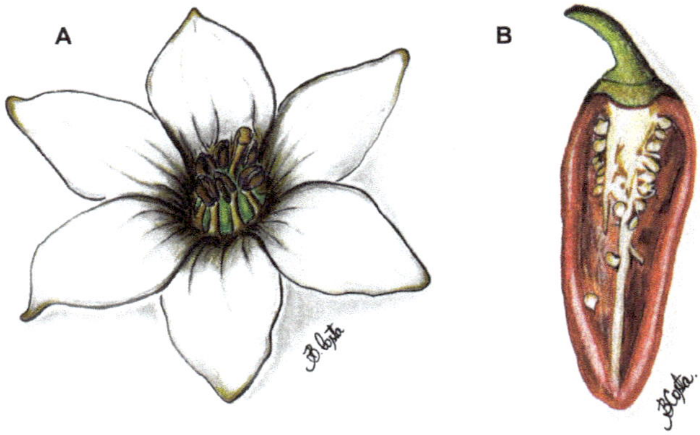

Figure 2.5 *Capsicum annuum* (A) flower and (B) fruit. Image credit: Barbara Elisabeth Teixeira Costa.

Originating from South and Central America, *C. annuum* contains pungent and non-pungent cultivars, accessions and varieties with more than 200 different common names. *C. annuum* can grow in dry, sub-arid, sub-humid and humid climates. It can be eaten *in natura* and as a prepared spice. It consist of cultivars such as jalapeño, serrano and cayenne (Figure 2.6). Sweet bell pepper (Figure 2.6E) *in natura* and paprika, the dried powdered fruit (Figure 2.6C), are the most common commercial presentations of *C. annuum*.[2,4]

There several genotypes of *C. annuum* with distinct external and internal features, including fruit shape, fruit surface, colour, length, width, size and placentation (Figure 2.6D and E).[11–13] The species also includes the large sweet bell-shaped peppers, known as bell peppers, which have almost no pungency.

C. annuum var. *annuum* L. fruits can be described by the following average characteristics:[11,12]

- length 12 cm
- width 19.85 mm
- cordate shape with flexuous surface
- colours varying from green to red
- hollow inside, with bilocular locule, axile placentation and 39 seeds.

2.2.2 *Capsicum chinense* Jacq.

The origin of the *C. chinense* species is thought to be the Amazon basin and Caribbean region, where it is broadly domesticated and cultivated.[4,14] The *C. chinense* species contains highly aromatic and pungent varieties. Some species are only aromatic, almost without any pungent sensation, such as the varieties murupi, pimenta de Cheiro and biquinho (Figure 2.7). *C. chinense* also includes the most pungent peppers in the world. Among its most famous varieties are the Trinidad Moruga scorpion, Scotch bonnet, bhut jolokia, Carolina Reaper and habanero chillies (Figure 2.8).

Figure 2.6 *Capsicum annuum* (A) unripe jalapeño, (B) serrano, (C) powdered paprika, (D) ripe jalapeño and (E) sweet bell pepper. Image credit: Carlos Magno da Silva Antonio.

Capsicum chinense possibly had the same common ancestor species as *C. annuum* and *C. frutescens*, thus its distinction from these species is crucial. In general terms, *C. chinense* can be differentiated from *C. frutescens* by the colour of the corolla and from *C. annuum* by the number of flowers per node (Table 2.1). *C. chinense* is a small shrub that grows up to 1.5 m tall. The flowers often occur as two to five per node and are pendant. The flowers and the fruit calyx are poorly toothed and the leaves are usually wrinkled (*e.g.* Carolina Reaper).[4,14] The fruit colours and shapes of *C. chinense* are even more diverse than those of *C. annuum*. For instance, its fruits can be yellow, orange, red or dark brown. The fruits have an average length of 47 mm and a width of 21 mm. The shapes and textures vary from long and smooth to triangular and wrinkled. There are also genotypes with rounded fruits.[4,15,16]

2.2.3 *Capsicum frutescens* L.

Capsicum frutescens is also known for its pungency and one of its cultivars, tabasco, is the main raw ingredient of the famous Tabasco sauce. Another major cultivar is malagueta (Figure 2.9).

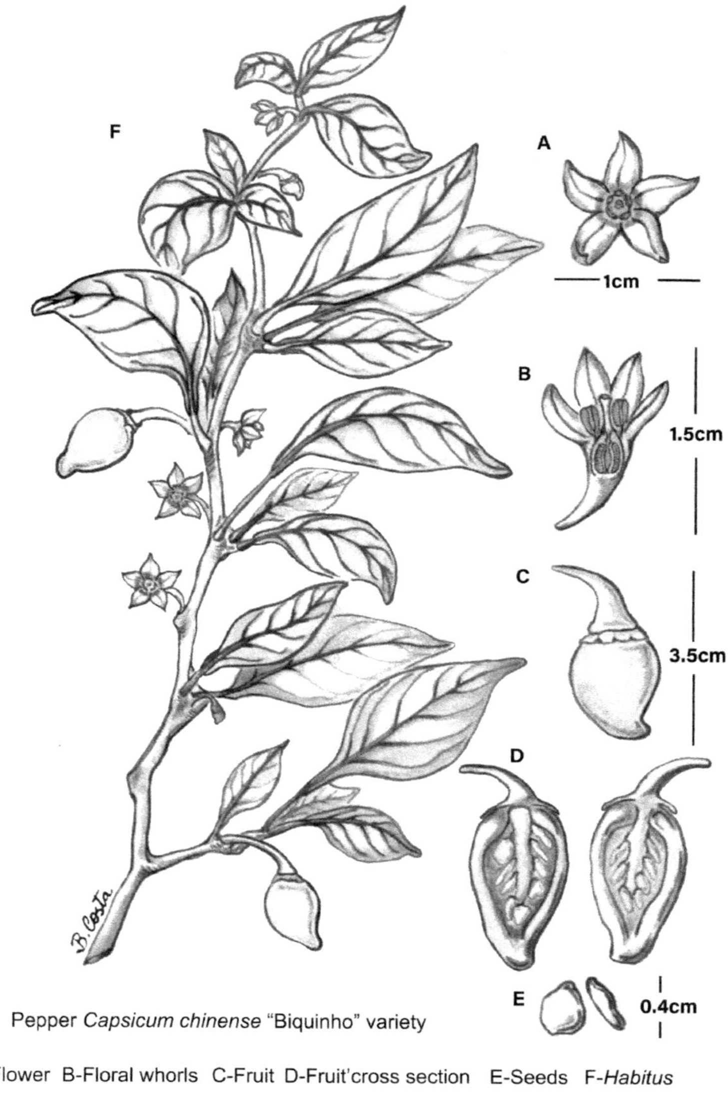

Figure 2.7 *Capsicum chinense* biquinho. Image credit: Barbara Elisabeth Teixeira Costa.

Capsicum frutescens is a small perennial shrub up to 2 m tall. One to three flowers occur per node with no calyx teeth. The corolla is greenish white with the petals folding back. The flower style and stigma can extend to at least 1.5 mm, passing the limit reached by the same structures in *C. chinense*. The fruits are usually conical and erect, significantly smaller than those of *C. chinense*, and with no teeth in the calyx. The average length and width of the fruits are 20 and 7 mm, respectively.[4,17,18]

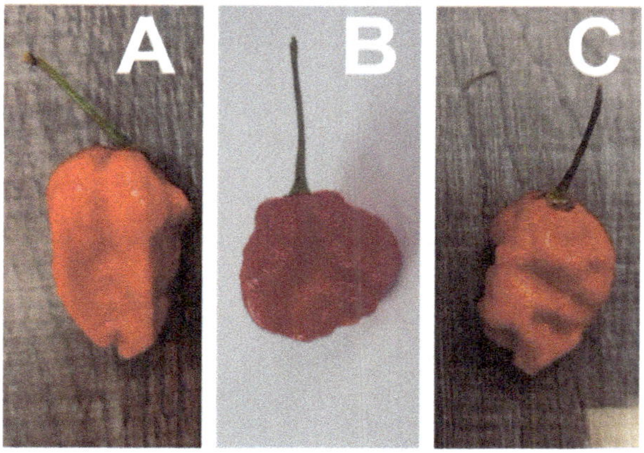

Figure 2.8 *Capsicum chinense* (A) bhut jolokia, (B) Carolina Reaper and (C) habanero. Image credit: Carlos Magno da Silva Antonio.

Figure 2.9 *Capsicum frutescens* (A) malagueta and (B) Tabasco sauce. Image credit: Carlos Magno da Silva Antonio.

Its geographical origin is thought to be the same as *C. chinense*, although most of its varieties are semi-domesticated or wild. *C. frutescens* can be cultivated in tropical and sub-tropical regions, but tabasco is the only economically reliable cultivar to be farmed outside Amazonia.

Malagueta is widely cultivated in Amazonia and became a popular regional spice.[17,18] The oleoresin of the malagueta pepper is prized for its

Table 2.2 Characteristics of the famous *Capsicum frutescens* cultivars.

Characteristic	Tabasco	Malagueta
Transitions of fruit colour during ripening	Light green to yellow to orange, then light red to dark red	Green to red
Length (cm)	2.5–5	1–3
Width (cm)	0.5	0.4–0.5
Pungency (SHU)	40 000	164 000

Figure 2.10 *Capsicum baccatum* var. cambuci. Image credit: Carlos Magno da Silva Antonio.

pungency (Table 2.2), sensory properties and bioactivity, making it a desired ingredient in the food industry.[19]

2.2.4 *Capsicum baccatum* var. *pendulum* (Willd.) Eshbaugh

Capsicum baccatum is a typical Latin American species and includes the varieties of aji, aji amarillo, campaña, cambuci (Figure 2.10) and cacho de cabra rojo. This species is an important ingredient in the traditional culinary practices of Peru and Bolivia, the region in which it was domesticated. The aspect of *C. baccatum* that differentiates it from the other domesticated *Capsicum* species is its resistance to disease. Such a characteristic makes this species valuable in the breeding of more resistant peppers.[20]

2.2.5 *Capsicum pubescens* Ruiz & Pavon

Capsicum pubescens is the least known of the domesticated *Capsicum* species. Its origin is thought to be near Bolivia and it was domesticated about

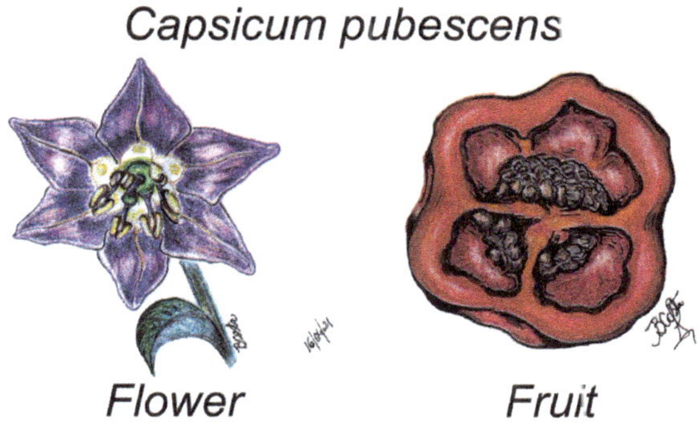

Figure 2.11 *Capsicum pubescens* var. rocoto. Image credits: Carlos Magno da Silva Antonio and Barbara Elisabeth Teixeira Costa.

6000 years ago. *C. pubescens* is mainly cultivated in Andean South America. Its most famous varieties include rocoto and canario (Figure 2.11). Morphological, chemical and genetic data for this species are scarce compared with other domesticated species.[21,22]

This species has unique morphological features. Its major differential features are hairy leaves and black seeds. *C. pubescens* is a prostrate shrub with a height of 30 cm and petioles up to 12 mm long. The flowers are purple with white spots in each node and a maximum of two flowers per node. The fruits can be dark green, red, yellow or brown, although it is more famous for its bright yellow colour, which inspired the name of the canario variety. The average length of *C. pubescens* fruits is 5 cm and they have roundish forms – for instance, manzano has a shape similar to an apple, canario is completely round and smooth like a tomato and peron is shaped like a pear.[4,21,23] Similar to *C. baccatum*, *C. pubescens* is also a valuable species for breeding varieties resistant to pathogens and suitable for cold climates.[4,23]

2.3 *Capsicum* Phylogenetics

The diversity of the colours, shapes, pungency and vegetative structures of *Capsicum* species suggests that correct identification is not a simple task, particularly for cultivars of the same species or the distinction of wild species. The correct identification of a plant is crucial when it is intended for human consumption and commercialisation. For instance, let us imagine that the Peruvian rocoto pepper (*C. pubescens*) could, at first glance, be mistaken for a tomato (*Solanum lycopersicum*) or a sweet bell pepper. The problem starts because the pungency in Scoville heat units (SHU) of a rocoto pepper is, on average, 65 000. It is not as hot as a habanero pepper, but can still be an unwelcome surprise. If the consumer is sensitive to capsaicinoids, then this could also be a health issue.[21]

The science of plant identification and classification is known as plant systematics. This field includes several identification methods, such as taxonomy, chemical profiling and genetic mapping. The major goals are to provide distinctions among botanical families and species, to characterise varieties and to determine evolutionary patterns. Knowing the evolutionary processes of plants helps us to better interact with flora and to develop methods to preserve and use plants more efficiently.

Most of the current developments in systematic botany are based on genetic information. This is because a plant's DNA contains all the information in its heritage and all that is going to be passed to future generations. It enables us to reconstruct the steps of evolution. The study of systematics by applying molecular sequence data is known as phylogenetics. In this research field, several sequences of DNA and RNA can be used to mark evolutionary patterns.

We can observe the diversity in *Capsicum* species with the naked eye, which reflects the wide genetic variability within this species. The domestication of *Capsicum* further increased this genetic diversity. The systematic classification of the genus becomes more difficult when we realise that we are unable to classify all the different species, varieties and cultivars based only on their morphological traits. There is a growing demand for more resistant varieties that can be cropped easily. In the beginning of the 2000s, the genetic exploration of the genus received a huge boost from the scientific community to help solve these two issues.[2,24]

Genetic mapping of the genus aimed to not only to systematically identify this species and to evaluate its evolutionary patterns in nature, but also to develop breeding programmes. The genetic map of *Capsicum* can be used to predict which species should be hybridised to obtain a specific characteristic, such as resistance to freezing conditions or the production of a certain SHU pungency.[5,25–27]

To analyse the phylogenetics of *Capsicum*, we must first look at its botanical family. The genus *Capsicum* is part of the Solanaceae family, which currently consists of 100 genera and nearly 2500 species. This family is considered to be a monophyletic group (a taxon that has the same common ancestor) and includes other important crop plants, such as potatoes, tomatoes and tobacco. Phylogenetic analysis of the Solanaceae family based on the chloroplast DNA regions *ndhF* and *trnLF* showed that *Capsicum* is part of the tribe Capsiceae, with a sister genus *Lycianthes*.[28] Even though the Solanaceae family has hundreds of species and *Capsicum* has a sister genus, *Capsicum* is the only genus to produce pungency. Genetically, pungency is generated by the locus *Pun1* and its candidate gene *AT3* (*Acyltransferase3*). *Pun1* is currently considered to be a molecular marker of the *Capsicum* genus and the only one that quantitively affects the concentration of capsaicinoids.[29,30]

The *Capsicum* genome is currently estimated to have about 37 989 scaffolds, 337 328 contigs and 34 903 genes, which are available in several gene databases. Within the *Capsicum* species, there are diploid and tetraploid species with 12 or 13 X chromosomes.[31] According to phylogenetic studies based on the molecular markers (a DNA fragment with a specific location in the genome) *matK*, *psbA-trnH* and *waxy*, the genus *Capsicum* is a monophyletic taxon originating from across the Andes of western to northwestern South America and contains

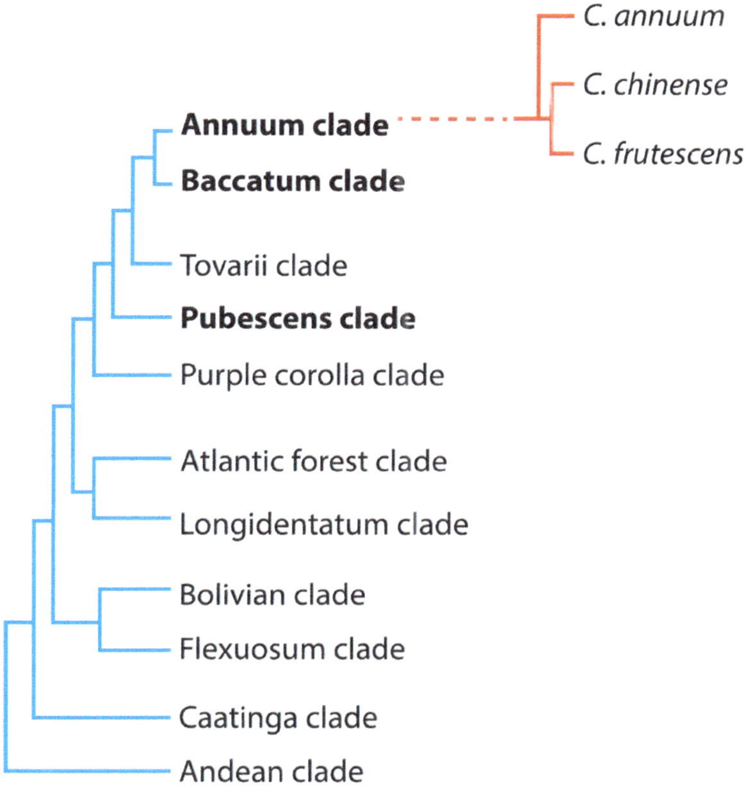

Figure 2.12 Simplified phylogenetic tree of *Capsicum* evolution. Image credit: Carlos Magno da Silva Antonio.

11 clades (Figure 2.12).[2,32] Because genetic traits control the expression of morphological and chemical features, it is possible to elucidate similarities in these organisation levels within the phylogenetic clades.

Phylogenetic evaluation of the genus highlighted that the pungency of *Capsicum* species and the greenish golden yellow fruits are evolutionary characteristics developed within the genus. The pungency can be associated with evolutionary processes that occurred after migration from the Andean region because most of the Andean clade species are non-pungent. By contrast, the red colour of chilli peppers is a plesiomorphic state, which means that it was inherited by a common ancestor of the Capsiceae tribe and therefore can also be found in the *Lycianthes* genus. The black colour of seeds is a characteristic of the Atlantic forest clade and also developed within the genus.[32]

2.4 Evolution and Breeding

The *Capsicum* genus contains a huge diversity of species, varieties and cultivars. This biodiversity is due to the genetic diversity, which is closely associated with the modern history of this genus through the strong speciation and genetic improvements obtained *via* breeding programmes. The plant speciation process refers to the process of evolution itself. Evolution enables species to adapt to specific environmental conditions by gene selection to perpetuate the species and usually occurs as a response to abiotic or biotic stress.

Speciation can occur *via* adaptive radiation, ecogeographical isolation, chromosomal rearrangements and hybridisation. All of these processes will result in a certain level of genetic flow, drift and recombination that can lead to the creation of a new lineage. Reproductive characteristics and genetic compatibility are among the features that will determine the preferential form of speciation.[31,33] In *Capsicum*, it is estimated that natural speciation occurs mainly by introgression (the recombination of a newer species, usually a hybrid, with its parental species) during geological events and climatic oscillations. The hybridisation of botanical species refers to the crossing of genetically distinct individuals, in which the genes of distinct species (*e.g. C. chinense* and *C. annuum*) are combined into a new species. This process can occur by natural means, but is more commonly induced by humans.

The natural diversity within *Capsicum* enabled human-directed hybridisation in several ways, although species compatibility is still a problem. As the demand for chilli peppers increased after the Age of Discovery, humans tried to increase the production of *Capsicum*. More crop fields were created in regions where *Capsicum* was not native and cultivation had to generate high yields. In nature, *Capsicum* species reproduce by both self-pollination and cross-pollination, with most of the expansion in range occurring *via* the dispersion of seeds by birds,[2,24,32] but these natural means of reproduction were unable to meet the new demands for chilli peppers. In addition, cultivating a species in a new location usually means that the species experiences a different climate.

Breeding programmes were therefore developed to crop large numbers of plants under distinct conditions. Breeding involves human-induced trait modifications in plants to obtain cultivars with desirable characteristics. There are several methods by which breeding programmes can be conducted with *Capsicum* (Figure 2.13), such as the in-field collection of particularly interesting individual plants or by mapping the genome to define the best species or varieties for cross-breeding. The major objectives in breeding programmes are to increase crop yields, enhance the sensory features of fruits, to increase resistance to abiotic stress, to increase resistance to diseases and pathogens, or to increase the content of particular chemicals (*e.g.* to increase the nutritional value or pungency). More than 290 genes are available to track the most desirable features.[25,33–35]

Among the genetic improvements that can be made in *Capsicum*, the most important are those related to disease resistance. *Capsicum* species are particularly susceptible to anthracnose disease caused by fungi of the *Colletotrichum* genus. This disease is characterised by dark spots on the leaves, followed by the death of vegetal tissue in these spots, in which the fungal

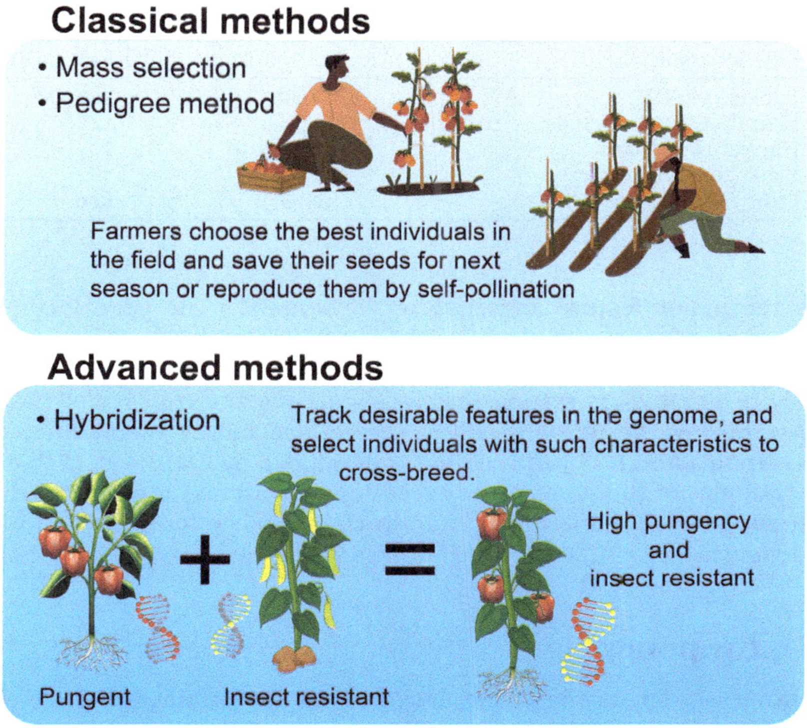

Figure 2.13 Examples of breeding programmes with *Capsicum*. Image credit: Carlos Magno da Silva Antonio.

spores are produced. Anthracnose is a serious issue in crops because it can drastically reduce production yields and compromise entire fields, not only of *Capsicum*, but also of strawberry, *Citrus* spp., sugarcane and maize. Genetic control to create accessions resistant to anthracnose is more healthy and less harmful to the environment than applying copper-based fungicides to the field. The genes used to improve disease resistance include:

- *Anr1, Anr2, Anr3* and *Anr4* for anthracnose disease
- *Bs1, Bs2, Bs3, Bs4, bs5, bs6* and *gds* for bacterial leaf spot
- *lmr-1, lmr-2* and *lmr-3* for powdery wilt
- *Me2, Me3, Me4* and *Me5* for resistance to *Meloidogyne* spp.
- *Mech1* and *Mech2* for root-knot nematodes
- *Tsw* for tospovirus tomato spotted wilt virus
- L^3, L^2, L^1 and L^* for tobacco mosaic virus
- *cm* for cucumber mosaic virus.

Regarding fruit quality, breeding programmes are also focused on the control of pungency. This characteristic is essential to *Capsicum* because

Table 2.3 Characteristics of cultivars developed in breeding programmes.

Characteristic	BRS Nandaia	BRS Juruti
Pungency (SHU)	200 000	260 000
Vitamin C (mg per 100 g of fruit)	124	122
Harvest time (days)	90	90
Yield (ton per hectare)	44	49
Fruit colour	Orange	Red

it is the major feature attractive to consumers. The gene locus that controls pungency was recognised in approximately 1935. It is known as *Pun1* or *C. Pun1* and is part of the BAHD acyltransferase family. This super family of enzymes is responsible for catalysing the transfer of the acyl moiety during the biosynthesis of several specialised metabolites. The BRS Nandaia and BRS Juruti cultivars developed by EMBRAPA (Table 2.3) are examples of the products of breeding programmes aiming to achieve high pungency, high levels of vitamin C, resistance to disease and high field yields.[36]

2.5 Cropping

Capsicum is cropped worldwide, but most of the farming areas are small (0.5–10 ha). In general, the genus requires warm temperate climates to develop and can be cultivated in different agricultural systems, such as open fields, agroforestry and greenhouse systems. Even small farms use methods to increase yields.[4,37]

Several factors can modify crop yields and must be rigorously controlled. For instance, the cultivation temperature should be maintained within the range 15–32 °C during the day and 15–18 °C during the night. Temperatures <15 °C can cause (depending on the cultivar) lower flowering rates and delayed plant development, which means smaller seedlings and specimens. Temperatures <5 °C can cause plant mortality. Peppers accept a wide range of soil types, but favour sandy loams with a pH in the range 5.5–6.8. Many people cultivate peppers indoors in small pots, although plants on farms are spaced at 60 cm × 70 cm.[37]

The crop cycle of peppers is usually 120–150 days depending on the cultivar and the growing conditions (Figure 2.14). *Capsicum* cultivars must be farmed without water-logging. Excessive water during cultivation can decrease the flowering rate of peppers. Their water requirements range from 600 to 900 mm during the crop cycle, although, if multiple harvesting is carried out, this can increase up to 1250 mm. Seeds for cropping can be purchased or produced by the farmer. Seedling growth is recommended to be performed in a nursery structure with rigorous control of the temperature, solar radiation, soil fertility and water supply. Transplantation to the field is recommended when the seedlings reach 15–20 cm in height; the average crop yield is 10 ton per ha.[37]

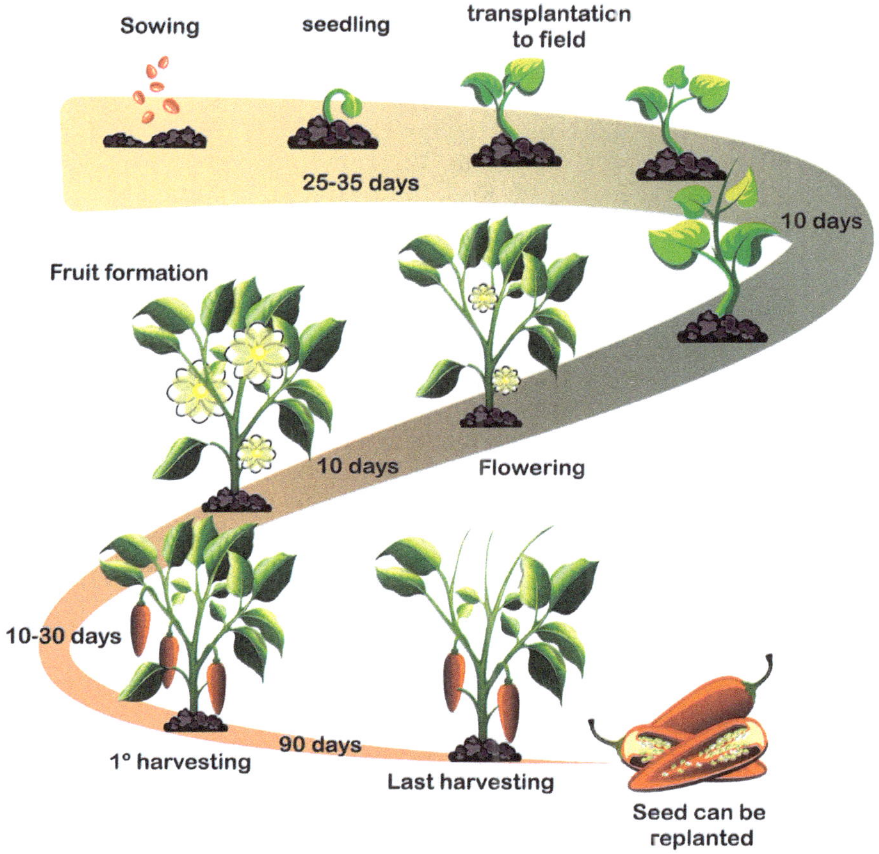

Figure 2.14 Crop cycle of *Capsicum* peppers. Image credit: Carlos Magno da Silva Antonio.

2.6 Concluding Remarks

Capsicum is a genus with many valuable resources, both economic and scientific. Its diversity of morphological features and its easy domestication have opened up opportunities to cultivate this genus worldwide. Different morphological traits, such as the appearance and taste of the fruits, are not just an attractive commercial feature, but are also used to distinguish between species, aiding the systematic classification of this genus and increasing our knowledge of the natural speciation process. *Capsicum* flowers are said to be complete because they have all the organs (*e.g.* stamens, petals and carpels) and can be recognised by particular colour variations, ranging from purple to dull white. Although these features are difficult to observe in an individual plant, they are nevertheless fundamental in the systematics of *Capsicum*. All these traits can also be observed in the genomics of *Capsicum*, which has one of the most well-described genomes of all crop plants. The genetic mapping of the genus is a crucial tool in its sustainable and ecological exploitation.

References

1. The Plant List 2013, http://www.theplantlist.org/tpl1.1/search?q=capsicum, accessed December 2020.
2. R. L. Jarret, G. E. Barboza, F. R. da, C. Batista, T. Berke, Y. Y. Chou, A. Hulse-Kemp, N. Ochoa-Alejo, P. Tripodi, A. Veres, C. C. García, G. Csilléry, Y. K. Huang, E. Kiss, Z. Kovacs, M. Kondrak, M. L. Arce-Rodriguez, M. A. Scaldaferro and A. Szoke, *J. Am. Soc. Hortic. Sci.*, 2019, **144**, 3–22.
3. G. E. Barboza, L. de Bem Bianchetti and J. R. Stehmann, *PhytoKeys*, 2020, **140**, 125–138.
4. V. M. Russo, *Peppers: botany, production and uses*, CABI International 2012, Wallingford, 2012.
5. B. M. Walsh and S. B. Hoot, *Int. J. Plant Sci.*, 2012, **162**, 1409–1418.
6. K. Chiou and C. A. Hastorf, *Capsicum spp. at the preceramic sites of Huaca Prieta and Paredones, Chicama Valley, Peru*, 2012.
7. C. K. Khoury, D. Carver, D. W. Barchenger, G. E. Barboza, M. van Zonneveld, R. Jarret, L. Bohs, M. Kantar, M. Uchanski, K. Mercer, G. P. Nabhan, P. W. Bosland and S. L. Greene, *Divers. Distrib.*, 2020, **26**, 209–225.
8. G. E. Barboza, C. C. García, M. Scaldaferro and L. Bohs, *PhytoKeys*, 2020, **167**, 13–29.
9. G. E. Barboza and L. D. B. Bianchetti, *Syst. Bot.*, 2005, **30**, 863–871.
10. Pimenta (Capsicum spp.) – *Embrapa Hortaliças*, https://sistemasdeproducao.cnptia.embrapa.br/FontesHTML/Pimenta/Pimenta_capsicum_spp/botanica.html, accessed November 2020.
11. D. A. Zhigila, A. A. Abdulrahaman, O. S. Kolawole and F. A. Oladele, *J. Bot.*, 2014, **2014**, 1–6.
12. H. Ballina-Gómez, L. Latournerie-Moreno, E. Ruiz-Sánchez, A. Pérez-Gutiérrez and G. Rosado-Lugo, *Chil. J. Agric. Res.*, 2013, **73**, 329–338.
13. S. D. Deka, *Electron. J. Plant Breed.*, 2017, **8**, 660–667.
14. R. L. Jarret and T. Berke, *HortScience*, 2008, **43**, 1694–1697.
15. C. I. Domenico, J. P. Coutinho, H. T. Godoy and A. M. de Melo, *Hortic. Bras.*, 2012, **30**, 466–472.
16. L. S. Muñoz-Ramírez, L. P. Peña-Yam, S. A. Avilóes-Viñas, A. Canto-Flick, A. A. Guzmóan-Antonio and N. Santana-Buzzy, *HortScience*, 2018, **53**, 1772–1775.
17. S. I. C. Carvalho, L. B. Bianchetti, C. F. Ragassi, C. S. C. Ribeiro, F. J. B. Reifschneider, G. S. C. Buso and F. G. Faleiro, *Genet. Mol. Res.*, 2017, **16**, 1–18.
18. S. Yamamoto and E. Nawata, *Tropics*, 2004, **14**, 111–121.
19. A. C. de Aguiar, J. F. Osorio-Tobón, L. P. S. Silva, G. F. Barbero and J. Martínez, *J. Supercrit. Fluids*, 2018, **133**, 86–93.
20. P. M. Eggink, Y. Tikunov, C. Maliepaard, J. P. Haanstra, H. De Rooij, A. Vogelaar, E. W. Gutteling, G. Freymark, A. G. Bovy and R. G. Visser, *Theor. Appl. Genet.*, 2014, **127**, 373–390.

21. S. W. Meckelmann, C. Jansen, D. W. Riegel, M. van Zonneveld, L. Ríos, K. Peña, E. Mueller-Seitz and M. Petz, *Eur. Food Res. Technol.*, 2015, **241**, 817–825.
22. S. Yamamoto, T. U. D. Jarwaningsih and H. A. W. Iriadinata, *Econ. Bot.*, 2013, **67**, 161–170.
23. G. Csillléry, *Acta Agron. Hungarica*, 2006, **54**, 151–166.
24. E. A. Moscone, M. A. Scaldaferro, M. Grabiele, N. M. Cecchini, Y. S. García, R. Jarret, J. R. Daviña, D. A. Ducasse, G. E. Barboza and F. Ehrendorfer, *Acta Hortic.*, 2007, **745**, 137–169.
25. C. Silvar and C. A. García-González, *Sci. Hortic.*, 2017, **218**, 249–257.
26. Y. Tanaka, H. Yoneda, M. Hosokawa, T. Miwa and S. Yazawa, *Sci. Hortic.*, 2014, **165**, 242–245.
27. M. J. Rodríguez-Maza, A. Garcés-Claver, S. W. Park, B. C. Kang and M. S. Arnedo-Andrés, *Mol. Breed.*, 2012, **30**, 889–898.
28. R. G. Olmstead, L. Bohs, H. A. Migid, E. Santiago-Valentin, V. F. García and S. M. Collier, *Taxon*, 2008, **57**, 1159–1181.
29. K. Han, H. J. Jeong, J. Sung, Y. S. Keum, M. C. Cho, J. H. Kim, J. K. Kwon, B. D. Kim and B. C. Kang, *Mol. Breed.*, 2013, **31**, 537–548.
30. A. N. Egan, S. Moore, G. M. Stellari, B. C. Kang and M. M. Jahn, *PLoS One*, 2019, **14**, 1–24.
31. S. Kim, M. Park, S. I. Yeom, Y. M. Kim, J. M. Lee, H. A. Lee, E. Seo, J. Choi, K. Cheong, K. T. Kim, K. Jung, G. W. Lee, S. K. Oh, C. Bae, S. B. Kim, H. Y. Lee, S. Y. Kim, M. S. Kim, B. C. Kang, Y. D. Jo, H. B. Yang, H. J. Jeong, W. H. Kang, J. K. Kwon, C. Shin, J. Y. Lim, J. H. Park, J. H. Huh, J. S. Kim, B. D. Kim, O. Cohen, I. Paran, M. C. Suh, S. B. Lee, Y. K. Kim, Y. Shin, S. J. Noh, J. Park, Y. S. Seo, S. Y. Kwon, H. A. Kim, J. M. Park, H. J. Kim, S. B. Choi, P. W. Bosland, G. Reeves, S. H. Jo, B. W. Lee, H. T. Cho, H. S. Choi, M. S. Lee, Y. Yu, Y. Do Choi, B. S. Park, A. Van Deynze, H. Ashrafi, T. Hill, W. T. Kim, H. S. Pai, H. K. Ahn, I. Yeam, J. J. Giovannoni, J. K. C. Rose, I. Sørensen, S. J. Lee, R. W. Kim, I. Y. Choi, B. S. Choi, J. S. Lim, Y. H. Lee and D. Choi, *Nat. Genet.*, 2014, **46**, 270–278.
32. C. Carrizo García, M. H. J. Barfuss, E. M. Sehr, G. E. Barboza, R. Samuel, E. A. Moscone and F. Ehrendorfer, *Ann. Bot.*, 2016, **118**, 35–51.
33. D. Sokona, Y. D. Niamoye, N. S. Paul, A. Olagorite, D. N. Aminata, G. T. Kadidiatou, T. T. R. Aissata, K. S. Riba and D. B. Daouleacute, *Afr. J. Agric. Res.*, 2013, **8**, 1108–1114.
34. H. Padilha and R. Barbieri, *Aust. J. Basic Appl. Sci.*, 2016, **10**, 148–154.
35. Y. Wahyuni, V. Stahl-Hermes, A. R. Ballester, R. C. H. de Vos, R. E. Voorrips, A. Maharijaya, J. Molthoff, M. V. Zamora, E. Sudarmonowati, A. C. M. Arisi, R. J. Bino and A. G. Bovy, *Mol. Breed.*, 2014, **33**, 503–518.
36. A. Srivastava and M. Mangal, in *The Capsicum Genome, Compendium of Plant Genomes*, ed. N. Ramchiary and C. Kole, Springer Cham, 2019, pp. 25–55.
37. Crop informations: Peppers, http://www.fao.org/land-water/databases-and-software/crop-information/pepper/en/, accessed November 2020.

CHAPTER 3

Capsaicinoids: The Origin of Pungency

3.1 Introduction

Capsicum species, popularly known as chilli peppers or, simply, peppers, are renowned worldwide for their peculiar characteristic of causing a hot sensation when eaten.[1] Peppers have gained more attention in recent years *via* the internet and social media as people have started challenging others to endure the burning sensation caused by a single bite of a very spicy pepper, such as Carolina Reaper, while filming and sharing the experience. Despite their popularity, few people are familiar with the term "pungency", which is attributed to peppers, and even fewer know what causes the sensation of heat so characteristic of peppers.

You may be asking yourself, "why should I know any of this?", but if you intend to research or develop bioproducts with peppers, it is crucial to understand the origin of their heat because this is due to one class of metabolites: the capsaicinoids.[1] Even if you do not intend to delve deeper into this class, you should be aware that, in addition to causing this hot sensation, these metabolites are also used in the treatment of disease and in the development of cosmetics.[2] And a word of warning, if you want to try an internet challenge or taste a very spicy pepper recipe, you may want to know how to stop the burning sensation because drinking water will not provide any relief.

3.2 Pungency and the Capsaicinoids

Pungency can be defined as the ability of a specific food to generate heat, burning or a spicy sensation when consumed. This characteristic is typically

Food Chemistry, Function and Analysis No. 37
Chemistry and Nutritional Effects of Capsicum
By Valdir Florencio da Veiga, Jr, Larissa Silveira Moreira Wiedemann, Claudio Pereira de Araujo, Jr and Ananda da Silva Antonio

Published by the Royal Society of Chemistry, www.rsc.org

observed in *Capsicum* species and is the main reason behind their commercial popularity because it enables their use as a seasoning to remove food insipidity.[1] The ability to generate pungency is correlated with the chemical composition of peppers because certain compounds can cause irritation in the terminal nerves responsible for pain and heat sensation (the transient receptor potential vanilloid-1, VR1). Within chilli peppers, this irritation is specifically caused by a set of substances known as capsaicinoids.[3,4]

Capsaicinoids are a group of alkaloids biosynthesised exclusively by the *Capsicum* genus. This chemical class is currently composed of about 25 naturally biosynthesised compounds and nearly 135 that are produced synthetically or semi-synthetically.[5] Despite the fact that capsaicinoids can only be found naturally in peppers, their economic importance is broader than simply a seasoning in the culinary industry. The capsaicinoids are also known for their bioactivity as painkillers, antioxidants, antitumor drugs and anticarcinogenic agents.[2,3]

In general, the pungency of a pepper increases as its capsaicinoid content increases. The capsaicinoid concentration in the fruit varies from about 0.003 to 1% of the total dry weight. Peppers can simultaneously produce a broad variety of capsaicinoids, including the compounds capsaicin and dihydrocapsaicin (Figure 3.1), which typically comprise 89–98% of the total capsaicinoid content depending on the *Capsicum* species and its cultivation conditions.[1,6,7] The pungency will depend not only on the total amount of capsaicinoids, but also on the relative concentration of each type of capsaicinoid present in the fruit because each compound has a distinct pungency level.[8–10]

3.3 Chemical Characteristics of Capsaicinoids

Capsaicinoids are classified as alkaloids and their chemical structure consists of carbon, oxygen and nitrogen atoms arranged in one aromatic ring known as the vanillyl group, as well as an amide and a carbon chain (Figure 3.2). Despite its simple chemical composition, the diversity of capsaicinoids produced within the *Capsicum* genus is caused by the different types of carbon (alkyl) chains present in the molecular structure. As the biosynthesis of this chemical class involves the condensation of a fatty acid

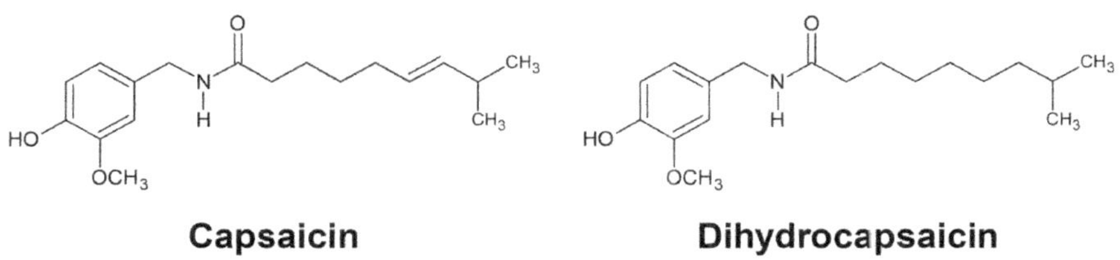

Figure 3.1 Chemical structures of major capsaicinoids produced by *Capsicum* species. Image credit: Carlos Magno da Silva Antonio.

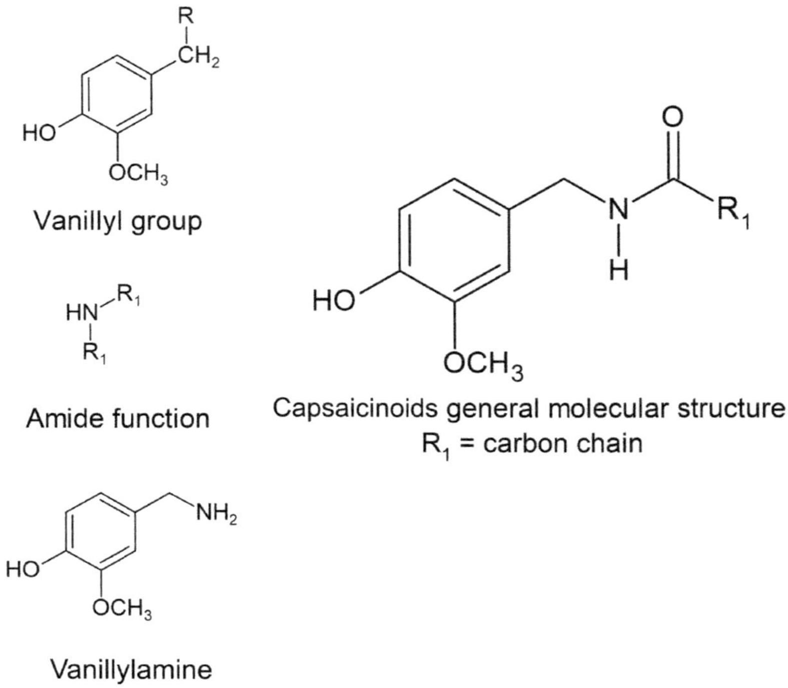

Figure 3.2 General structure of capsaicinoids and their main feature: the vanillyl group. Image credit: Carlos Magno da Silva Antonio.

with vanillylamine (Figure 3.2), each distinct fatty acid involved in the biosynthesis will produce a different capsaicinoid.[1,7,11]

Considering that the difference among capsaicinoids is due to the alkyl chain, it is natural to expect that the sub-classification of capsaicinoids into groups is based on the characteristics of their carbon chain. Capsaicinoids can be classified into three major groups (Figure 3.3) based on whether their alkyl chain is unsaturated and ramified (the capsaicin group), saturated and ramified (the dihydrocapsaicin group) or saturated and linear (the *N*-vanillyl-*n*-acylamide group).[12,13]

It is easy to observe in the general structure of capsaicinoids (Figure 3.2) that they present amphiphilic properties. This characteristic is caused by the presence of three distinct regions within the molecular structure: (1) the hydrophilic region comprising the vanillyl group; (2) a dipolar region comprising the amide in the middle of the structure; and (3) a lipophilic region comprising the carbon chain. The amphiphilic behaviour of capsaicinoids is fundamental in their ability to produce pungency because this property enables the capsaicinoids to interact with the terminal nerves responsible for pain and heat sensation (VR1) and with cell membranes without degradation of their chemical structure.[1,3,7,14] In addition, because several of the

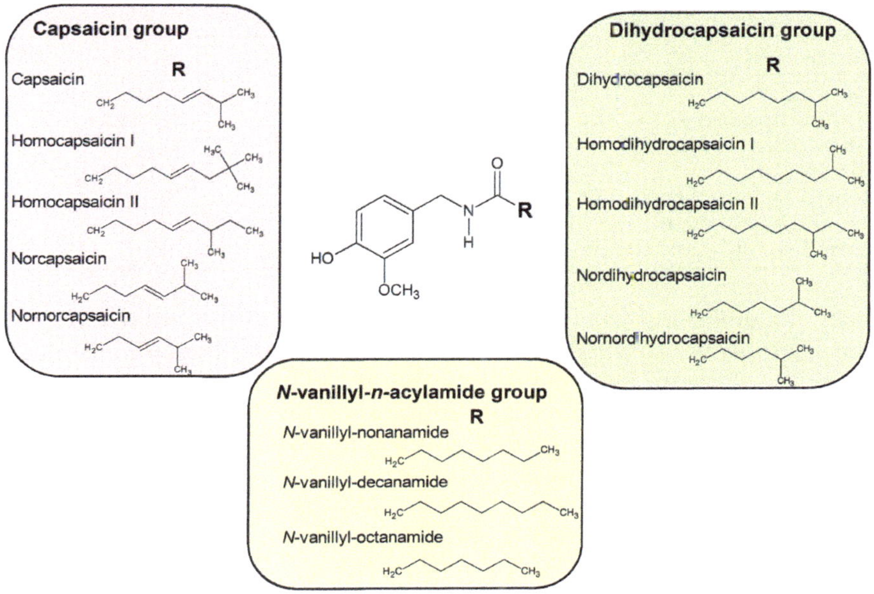

Figure 3.3 Major compounds present in the three groups of capsaicinoids. Image credit: Carlos Magno da Silva Antonio.

alkaloids contain an amide in their structure, the capsaicinoids present no basic behaviour.[15]

Because terminal nerves similar to VR1 and cell membranes have sites that can interact with proton species (known as proton channels), such as Ca^{2+}, the interaction of capsaicinoids with them can be mainly attributed to the vanillyl moiety (the hydrophilic and dipolar region of the molecule). The vanillyl moiety acts as a proton donor, allowing it to attach to the proton channels.[2,12] As proton donors, the capsaicinoids can also stabilise radical species. Unstable radicals within plants and animals are responsible for the oxidative stress that causes cell death, ageing and diseases such as cancer. Capsaicinoids therefore act as antioxidants, a fact that has encouraged their use in pharmacology despite their pungency.[2,3,16]

Capsaicinoids are weakly resistant to thermal energy, with melting points ranging from 40 to 80 °C (Table 3.1) and boiling points around 210 °C (capsaicin). In general, at least 50% of the original concentration of most capsaicinoids decomposes at temperatures >80 °C when the carbon–nitrogen bond in the amide is cleaved, producing a vanillyl moiety and a fatty acid.[1,12,17]

Despite their amphiphilic behaviour, capsaicinoids have a low solubility in water. Their solubility is higher in medium- to low-polarity solvents, such as alcohols.[12] This low solubility in water is also why you will not be able to reduce the heat sensation with water after eating a chilli pepper. A low-polarity liquid must be used instead, such as an oil or milk, because this will

Table 3.1 Molecular mass and melting point of selected capsaicinoids.

Capsaicinoid	Exact molecular mass (g mol^{-1})	Melting point (°C)
Capsaicin	305.199094	62–65
Dihydrocapsaicin	307.214744	65.5–65.8
Homocapsaicin I	319.214744	64.5–65.5
Norcapsaicin	291.183444	42.5–44.0
Homodihydrocapsaicin I	321.230394	70–71
Nordihydrocapsaicin	293.199094	60–61
N-Vanillyl nonivamide	293.199094	54

readily solubilise the capsaicinoids and remove them from nerve terminals. The solubility of capsaicinoids is a fundamental consideration in extraction procedures because it will determine the efficacy of the method and therefore the final capsaicinoid concentration in the bioproduct. Methanol, ethanol, ethyl acetate, acetonitrile, acetone and dichloromethane could all be used for the extraction of capsaicinoids. Most of these solvents are highly toxic to humans and therefore the food industry tends to use ethanol because it is the least aggressive and the easiest to remove from the final product.[12,18,19]

3.4 Pepper Pungency

There are several types of chilli pepper with different pungencies or "burning" levels. The pungency depends on the total amount of capsaicinoids within the pepper. It is important to remember that the total capsaicinoid concentration can vary from 0.003 to 1% of the total dry fruit weight. However, just because a pepper is pungent does not mean that a single fruit contains several grams of capsaicinoids. The concentrations are in the milligram range, emphasising the efficacy of their interaction with the terminal nerves in generating the sensation of heat.[3,12,20,21]

The pungency of peppers is currently desirable, not only due to the beneficial health effects of capsaicinoids, but also to their use in culinary products.[22] Several different methods are used to compare the heat of peppers; many of these are instrumental methods, but the most famous is an organoleptic test that only requires a well-trained person.[12] This test is very subjective to the tester's opinion, but is still used to scale the heat of peppers.[8,23]

This organoleptic test is known as the Scoville heat test, named after its creator Wilbur Scoville. Created in 1912, this test is based on how much of a water–sugar solution is needed to dilute the pepper until a point at which it is not possible to feel its pungency. For example, if one cup of chilli pepper sauce requires 5000 cups of a water–sugar solution to completely remove the perception of hotness, then its pungency according to the Scoville heat test is 5000 Scoville heat units (SHU).[24,25] In other words, the SHUs measure how many parts per million a pepper must be diluted by to not produce any burning sensation (Figure 3.4).

Among the diversity of cultivated peppers, three have already entered the *Guinness Book of World Records* as the "hottest peppers in the world": the bhut jolokia in 2007 with 1 001 304 SHU; the Trinidad Moruga scorpion in 2012 with 2 009 231 SHU; and the Carolina Reaper in 2013 with 2 200 000 SHU.[8,26] For a sensible person, even licking a toothpick used to pierce a Carolina Reaper can cause an intense burning sensation. To illustrate what this means, pure capsaicin is 16 000 000 SHU, oven-dried paprika is 20 728 SHU and jalapeño pepper is 21 700 SHU (Figure 3.5).

Although capsaicin is, in general terms, the major capsaicinoid within chilli peppers and therefore the main compound responsible for the pungency, the content of other capsaicinoids cannot be overlooked. Some varieties of peppers (*e.g.* omnicolor, jalapeño, puya and serrano) contain more dihydrocapsaicin than capsaicin (Table 3.2).

Although capsaicin is not their major capsaicinoid, these peppers still maintain a medium pungency because dihydrocapsaicin is about 15 000 000 SHU, whereas other minor capsaicinoids, such as homocapsaicin and dihydrocapsaicin, are 8 600 000 and 9 100 000 SHU, respectively.[27–29]

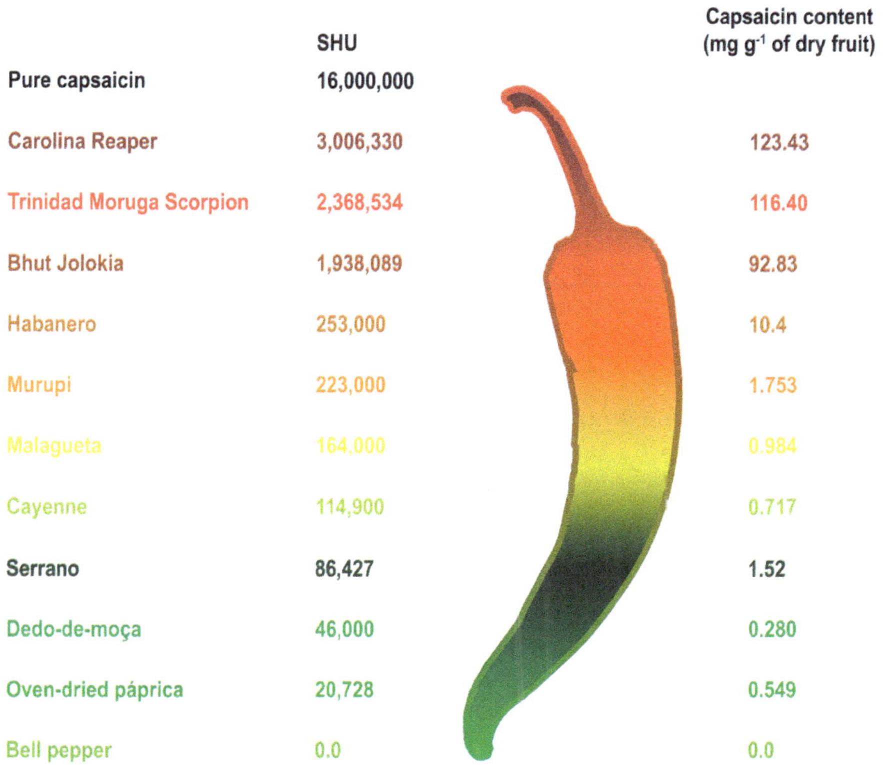

Figure 3.4 Scoville heat units (SHU) and capsaicin content of popular types of pepper. Image credit: Carlos Magno da Silva Antonio.

Figure 3.5 Carolina Reaper and Trinidad Moruga scorpion high-pungency peppers. Image credit: Carlos Magno da Silva Antonio.

Table 3.2 Capsaicinoid profile ($\mu g\, g^{-1}$ of dry fruit) in peppers with a low capsaicin content.[a]

Capsicum variety	C	DHC	HC	NDHC	SHU
C. annuum piquillo[b]	1.35	0.24	—	—	500
C. annuum padrón[b]	553.47	166.96	—	—	2500
C. annuum Alegría[b]	766.26	269.44	—	—	50 000
C. chinense naga morich	37 510	13 592	—	—	1 500 000
C. chinense Carolina Reaper	131 760	68 470	—	—	3 006 330
C. chinense Trinidad scorpion	98 120	34 440	—	—	2 368 534
C. baccatum ají omnicolor	4.71	6.66	1.79	0.72	62 000
C. annuum jalapeño cv. Don Julio	8030	9390	160	2480	303 602
C. annuum puya	1180	2320	80	550	61 526
C. annuum serrano cv. Don Diego	1520	3540	0.0	530	86 427

[a]C = capsaicin; DHC = dihydrocapsaicin; HC = homocapsaicin; NDHC = dihydrocapsaicin; SHU = Scoville heat unit.
[b]$\mu g\, g^{-1}$ of fresh fruit.

3.5 Biological Function and Metabolism in Plants

The production of capsaicinoids in plants is related to their need to adapt to certain environments.[21] Plants change their metabolism to produce compounds that enable their survival and reproduction in a certain biome. The capsaicinoids are a defence tool used by several *Capsicum* species to avoid attacks by small mammals (*e.g.* squirrels) and to favour their consumption by birds. Unlike mammals, birds are insensitive to pungency and their digestive systems do not change the reproductive capacities of the seed.[21,30] They can therefore help the plant species to proliferate by spreading the seeds in their faeces. In addition, the presence of capsaicinoids in seeds prevents the growth of fungi, which could harm the plant.[30] Not all *Capsicum* species can produce capsaicinoids, however. These peppers, known as "non-pungent", are mainly related to ancient species that underwent fewer adaptative processes.[31–33]

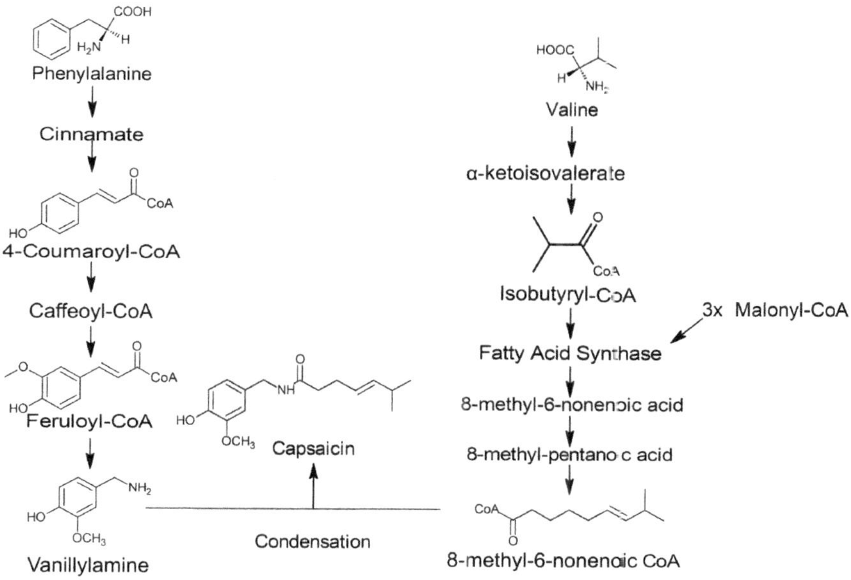

Figure 3.6 Summary of the biosynthetic pathway of capsaicinoids. Image credit: Carlos Magno da Silva Antonio.

The biosynthesis of capsaicinoids mainly occurs in the epidermal cells of the pepper's placenta and can be summarised as a condensation reaction between the vanillylamine and a fatty acid through several enzymatic reactions. The biosynthesis of capsaicinoids involves the phenylpropanoid and branched-chain fatty acid pathways (Figure 3.6).[34] Capsaicinoids mainly form in the placenta because there are high concentrations of vanillylamine in this organ. The vanillylamine is derived from phenylalanine, the main precursor of phenylpropanoid biosynthesis and therefore also the primary precursor of capsaicinoids. By contrast, the acyl moiety of capsaicinoids can be derived from valine or leucine.[1,7]

3.6 Accumulation of Capsaicinoids

The biosynthesis of capsaicinoids is most intense in the first 50 days after flowering. The fruit is consequently more pungent within this time period. The reduction of pungency occurs slowly as the concentration of capsaicinoids decreases after this period. Capsaicinoids are widespread in both the fruit and the vegetative organs, such as the leaves, although the highest concentration of capsaicinoids is in the placenta. It is estimated that the amount of capsaicinoids within the fruit can be 10–33 larger in the placenta than in other parts, such as the seeds and the pericarp (Figure 3.7).[1,35]

The pungency of chilli peppers is increased by the capsaicinoid content and this is the main reason why consumers are recommended to remove the

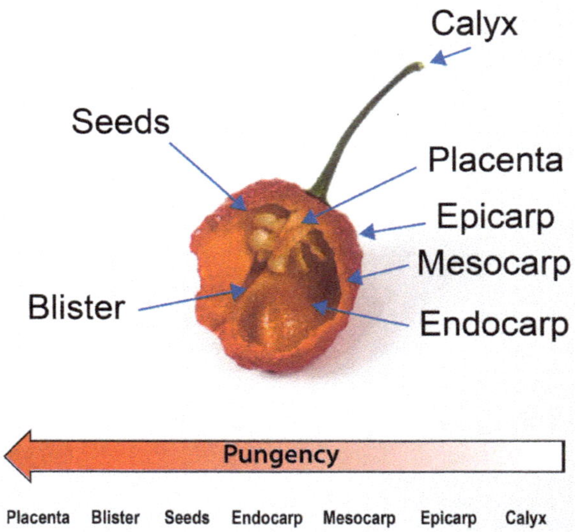

Figure 3.7 General pungency of each part of a chilli pepper. Image credit: Carlos Magno da Silva Antonio.

inner white part of the pepper (the placenta) before eating if they do not want to experience too much heat. Of course, there are always exceptions in nature and extremely hot peppers, such as the bhut jolokia and Trinidad Moruga scorpion (both *Capsicum chinense* species) also have a large accumulation of capsaicinoids in their pericarp.[6]

The production and accumulation of capsaicinoids within each chilli is influenced by several internal and external factors. The internal (endogenous) factors are mainly related to the DNA code of the plant. The biosynthesis of capsaicinoids is regulated by a specific allele in DNA, *Pun1*, which can be up- or down-regulated to produce higher or lower amounts of capsaicinoids depending on the species, cultivar or even plant for adaptation in a new environment. For example, because capsaicinoids have a role in the defensive system of plants against insects, *Pun1* can be up-regulated and produce more capsaicinoids in a stressful environment in which the plant is subjected to more predators.[1,36]

Several environmental (exogenous) factors can influence the accumulation of capsaicinoids. These include, but are not limited to, the environmental temperature, the incidence of solar radiation and the soil fertility. Despite intense research efforts to understand the phytochemistry of chilli peppers, there is still contradicting data about how each exogenous factor influences the accumulation of capsaicinoids. Most of the existing data on this topic usually correlate the effect to a specific cultivar, variety or species because each of these responds differently.

Understanding the pattern of accumulation of capsaicinoids in *Capsicum* can aid research, such as optimisation of the domestication of specific varieties, systematic evaluation (the distinction of species though their chemical profile) and improvements in cultivation. For instance, the cultivation of chilli peppers for consumption as a pungent fruit or food additive may use a specific water regime to increase the capsaicinoid content. Highly pungent cultivars, such as bhut jolokia and orange habanero, increase their capsaicinoid concentration in environments with a low water supply (only receiving water every two days).[37]

Another factor affecting the capsaicinoid content is the stage of ripening. The modification of the capsaicinoid content during fruit maturation is related to their role as protective metabolites and in the pollination process. The pungency of capsaicinoids repels noxious insects to protect the plant during growth and pollination and to protect it against oxidative stress from solar radiation. In general, the time at which peppers will have the highest capsaicinoid concentration is when the fruit reaches maturity, usually 40–50 days after fruiting.[1,36,37]

3.7 Non-pungent Peppers

Chilli peppers have several commercial applications, not only as a result of their taste, but also due to their bioactivity against many diseases. A pertinent issue in the application of such exotic fruit for pharmacological purposes is their pungency and the toxic effects that can be caused by small quantities of capsaicinoids. A solution to this is the evaluation of non-pungent species and their capsaicinoids and capsinoids.

Non-pungent peppers are cultivars in which the *Pun1* allele is nearly or completely silenced, meaning that these species are unable to produce capsaicinoids, although they still continue to synthesise non-pungent capsaicinoid analogues known as capsinoids. The major difference between capsinoids and capsaicinoids is that capsinoids have an ester instead of an amine, which makes their structure more hydrophilic. The absence of the amide function is responsible for the absence of pungency. Pungent and non-pungent species, varieties and cultivars of *Capsicum* are able to synthesise capsinoids, however, the pungent species only produce low concentrations (Table 3.3). Capsinoids were first isolated from a cultivar of *Capsicum annuum* named CH-19 Sweet. Major capsinoids are capsiate and dihydrocapsiate (Figure 3.8).[1,7]

Table 3.3 Capsiate concentration in *Capsicum* species.

Species, variety or cultivar	Capsiate content	
Red habanero (*C. chinense*)	156.97	$\mu g\,g^{-1}$ of fruit dry weight
Bhut jolokia (*C. chinense*)	440.90	$\mu g\,g^{-1}$ of fruit dry weight
Chile de arbol (*C. annuum*)	5.00	$\mu g\,g^{-1}$ of fruit dry weight
Tabasco (*C. frustescens*)	20.62	$\mu g\,g^{-1}$ of fruit dry weight
509-45-1 (*C. annuum*)	599	$\mu g\,g^{-1}$ of fruit weight
C. annuum var. glabriusculum	189.7	$\mu g\,g^{-1}$ of fruit weight
Malagueta (*C. frutescens*)	431.10	$\mu g\,g^{-1}$ of fruit weight

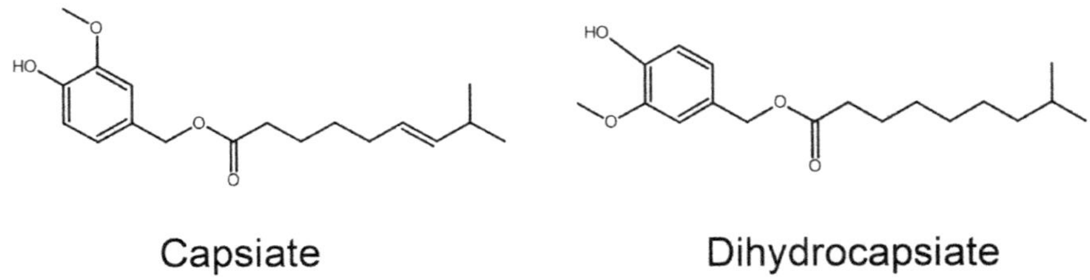

Figure 3.8 Chemical structures of the major capsinoids in sweet peppers. Image credit: Carlos Magno da Silva Antonio.

Capsinoid bioactivity is like that of capsaicinoids and the absence of pungency has made them better candidates in cancer treatment and weight loss. Their biological activity includes pain relief, cardiovascular protection, thermogenic effects and non-insulin blood sugar control.[1,36]

3.8 Concluding Remarks

The main chemical feature of chilli peppers is the alkaloids known as capsaicinoids. Their pungency is unique and characteristic of each species, variety and cultivar. Capsaicinoids can be controlled through modification not only of their genetic code, but also by environmental conditions such as the availability of water. Understanding the features of this class of chemicals is crucial in pharmacology and the food industry and can aid in pharmacological screening and the improvement of cultivation methods.

References

1. A. S. Antonio, L. S. M. Wiedemann and V. F. Veiga Junior, *RSC Adv.*, 2018, **8**, 25767–25784.
2. N. Baenas, M. Belovi, N. Ilic, D. A. Moreno and C. García-viguera, *Food Chem.*, 2019, **274**, 872–885.
3. M. Hayman and P. C. A. Kam, *Curr. Anaesth. Crit. Care*, 2008, **19**, 338–343.
4. Y. Wahyuni, A. R. Ballester, E. Sudarmonowati, R. J. Bino and A. G. Bovy, *J. Nat. Prod.*, 2013, **76**, 783–793.
5. V. Gómez-Calvario, M. L. Garduño-Ramírez, I. León-Rivera and M. Y. Rios, *Magn. Reson. Chem.*, 2016, **54**, 268–290.
6. R. Sugiyama, *Cytologia*, 2017, **82**, 303–306.
7. M. L. Arce-Rodríguez and N. Ochoa-Alejo, *Plant Cell Rep.*, 2019, **38**, 1017–1030.
8. L. S. Muñoz-Ramírez, L. P. Peña-Yam, S. A. Avilóes-Viñas, A. Canto-Flick, A. A. Guzmóan-Antonio and N. Santana-Buzzy, *HortScience*, 2018, **53**, 1772–1775.

9. D. Giuffrida, P. Dugo, G. Torre, C. Bignardi, A. Cavazza, C. Corradini and G. Dugo, *Food Chem.*, 2013, **140**, 794–802.
10. O. Monago-Maraña, M. Guzmán-Becerra, A. Muñoz de la Peña and T. Galeano-Díaz, *J. Food Compos. Anal.*, 2018, **67**, 10–18.
11. J. Díaz, F. Pomar, Á. Bernal and F. Merino, *Phytochem. Rev.*, 2004, **3**, 141–157.
12. A. da Silva Antonio, L. S. M. Wiedemann and V. F. da Veiga Junior, *Food Anal. Methods*, 2019, **12**, 1327–1345.
13. U. Schweiggert, R. Carle and A. Schieber, *Anal. Chim. Acta*, 2006, **557**, 236–244.
14. X. J. Luo, J. Peng and Y. J. Li, *Eur. J. Pharmacol.*, 2011, **650**, 1–7.
15. E. Fattourusso and O. Taglialatela-Scafati, *Modern Alkaloids: Structure, Isolation, Synthesis and Biology*, Wiley-VCH, Germany, 2nd edn, 2007.
16. N. Riquelme and S. Matiacevich, *CyTA-J. Food*, 2017, **15**, 344–351.
17. W. Si, S. W. Man, Z. Y. Chen and H. Y. Chung, *Nat. Prod. Commun.*, 2014, **9**, 985–988.
18. R.-M. Stoica, M. Moscovici, C. Tomulescu and N. Babeanu, *Sci. Bull. Ser. F. Biotechnol.*, 2016, **20**, 93–98.
19. L. Koleva Gudeva, V. Maksimova, M. Serafimovska Darkovska, R. Gulaboski and E. Ivanovska Janevik, *Sci. Work. Food Sci. Eng. Technol.*, 2013, **60**, 917–922.
20. Y. J. Surh and S. Sup Lee, *Life Sci.*, 1995, **56**, 1845–1855.
21. E. R. Naves, L. de Ávila Silva, R. Sulpice, W. L. Araújo, A. Nunes-Nesi, L. E. P. Peres and A. Zsögön, *Trends Plant Sci.*, 2019, **24**, 109–120.
22. N. Baenas, M. Belović, N. Ilic, D. A. Moreno and C. García-Viguera, *Food Chem.*, 2019, **274**, 872–885.
23. S. Jackson, D. J. Swiner, P. C. Capone and A. K. Badu-Tawiah, *Anal. Chim. Acta*, 2018, **1023**, 81–88.
24. P. H. Todd, M. G. Bensinger and T. Biftu, *J. Food Sci.*, 1977, **42**, 660–665.
25. J. D. Batchelor and B. T. Jones, *J. Chem. Educ.*, 2000, **77**, 266.
26. P. W. Bosland and J. B. Baral, *HortScience*, 2007, **42**, 222–224.
27. A. S. Antonio, G. D. Fonseca, Á. J. B. Mendes, L. S. M. Wiedemann and V. F. Veiga-Junior, *J. Braz. Chem. Soc.*, 2019, **30**, 1115–1129.
28. C. Zhou, D. Ma, W. Cao, H. Shi and Y. Jiang, *Food Addit. Contam., Part A*, 2018, **35**, 1447–1452.
29. A. C. De Aguiar, J. P. Coutinho, G. F. Barbero, H. T. Godoy and J. Martínez, *Int. J. Food Prop.*, 2016, **19**, 1292–1302.
30. G. Csilléry, *Acta Agron. Hung.*, 2006, **54**, 151–166.
31. D. Y. Kwon, Y. S. Kim, S. Y. Ryu, M. R. Cha, G. H. Yon, H. J. Yang, M. J. Kim, S. Kang and S. Park, *J. Nutr. Biochem.*, 2013, **24**, 1078–1085.
32. E. Blum, M. Mazourek, M. O'Connell, J. Curry, T. Thorup, K. Liu, M. Jahn and I. Paran, *Theor. Appl. Genet.*, 2003, **108**, 79–86.
33. C. Aza-González, H. G. Núñez-Palenius and N. Ochoa-Alejo, *Plant Cell Rep.*, 2011, **30**, 695–706.

34. M. L. Arce-Rodríguez and N. Ochoa-Alejo, *Plant Physiol.*, 2017, **174**, 1359–1370.
35. S. Taira, S. Shimma, I. Osaka, D. Kaneko, Y. Ichiyanagi, R. Ikeda, Y. Konishi-Kawamura, S. Zhu, K. Tsuneyama and K. Komatsu, *Int. J. Biotechnol. Wellness Ind.*, 2012, **1**, 61–65.
36. V. G. Uarrota, M. Maraschin, â. D. F. M. de Bairros and R. Pedreschi, *Crit. Rev. Food Sci. Nutr.*, 2021, **61**, 649–665.
37. N. Jeeatid, S. Techawongstien, B. Suriharn, S. Chanthai, P. W. Bosland and S. Techawongstien, *Food Chem.*, 2018, **245**, 792–797.

CHAPTER 4

Specialised Bioactive Metabolites of Capsicum

4.1 Introduction

Peppers shall not live on capsaicinoids alone. The chemical composition of chilli peppers is, in fact, wide and diverse, as with any other plant. Specialised metabolites are essential to plant survival, growth and adaptation to specific environments. It is estimated that there are >100 000 distinct metabolites within the Kingdom Plantae. Several chemical classes can be encountered within the *Capsicum* genus, including carotenoids, polyphenols, fatty acids, terpenoids and volatile organic compounds (VOCs).[1]

Each chemical class has its role in the fruit of *Capsicum* and is produced in a distinct pattern. If you think about chilli peppers beyond their pungency, it is easy to note other characteristics, such as aroma, taste, strong colours and the anaesthetic-like sensation after eating one. These are just the perceivable features promoted by the combination of specialised metabolites in *Capsicum*. For instance, carotenoids and flavonoids are the main compounds responsible for the variety of observed colours, whereas volatile organic acids generate distinct aromas.

These specialised metabolites result in *Capsicum* fruits presenting other characteristics that are not apparent while eating. Polyphenols, carotenoids and terpenoids also confer antioxidant, anti-inflammatory, antidiabetic and antimicrobial properties and can act as painkillers and cardiovascular protectants. *Capsicum* fruits also have nutritional relevance because they have high concentrations of vitamins C and E, which are essential in the human immune system.[2,3]

Food Chemistry, Function and Analysis No. 37
Chemistry and Nutritional Effects of Capsicum
By Valdir Florencio da Veiga, Jr, Larissa Silveira Moreira Wiedemann, Claudio Pereira de Araujo, Jr and Ananda da Silva Antonio

Published by the Royal Society of Chemistry, www.rsc.org

A close look in a market or pharmacy will find a huge variety of products listing chilli peppers among their ingredients, including dyes, spices, vitamin supplements, and natural topical and oral drugs used as anti-inflammatories. For each type of product there is a specific combination of chemical compounds that are suitable for the task, but not every chilli pepper has the same chemical composition. Which chilli peppers can be exploited for medicines and which are more suitable as food dyes? How do the flavonoids accumulate during ripening? Which carotenoids are commonly found in *Capsicum*? How does abiotic stress influence the biosynthesis of specialised metabolites? These are fundamental questions when developing new products with this plant group.

4.2 Carotenoids

The carotenoids are a group of terpenoids containing five units of isoprene. Classified as tetraterpenoids, this chemical class consists of nearly 750 compounds. Carotenoids are widely distributed and naturally biosynthesised by plants, having an important role in photosynthesis, the prevention of oxidative processes and photoprotection. The variety of colours we observe in peppers are, similar to many other species, partially promoted by carotenoids.[1,4] Carotenoids can be classified by their molecular structure or by the colour they produce (Figure 4.1). The classification by molecular structure divides this chemical class into two groups: (1) carotenes, which are linear or cyclic structures containing only hydrogen and carbon atoms, such as carotene; and (2) xanthophylls, which are structures that containing oxygen atoms that form functional groups, such as hydroxyl, methoxyl and epoxy (*e.g.* violaxanthin).

A remarkable feature in the chemical structure of carotenoids is the alternating double bonds along the structure (Figure 4.1). This enables the delocalisation of electrons through the carbon chain, which gives carotenoids the ability to interact with visible light and present chromogenic properties. The wavelength with which the carotenoids interact also depends on other structural characteristics, such as the number and type of oxygenated moieties, the length of the carbon chain and the presence of cyclic ends.[5]

The classification of carotenoids by the colour they produce can also divide them into two groups (Figure 4.1): (1) a yellow fraction, with less oxidated carotenoids, mainly carotenes, hydroxylated and epoxidised carotenoids; and (2) a red fraction with a 3-hydroxyacyclopentane ring that is biosynthesised from the yellow fraction carotenoids.

4.2.1 Carotenoids of *Capsicum*

Capsicum species, both pungent and non-pungent, are considered to be rich sources of carotenoids. In addition to capsaicinoids, the carotenoids are also commercially interesting because this chemical class is remarkably

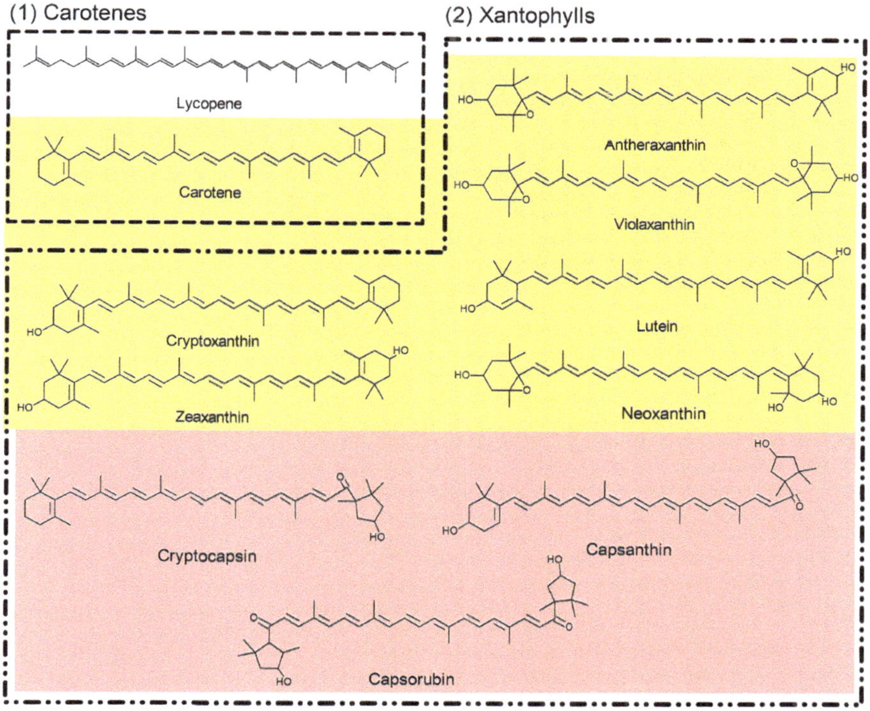

Figure 4.1 Chemical structures of carotenoids classified as carotenes, xanthophylls, and yellow and red fractions. Image credit: Carlos Magno da Silva Antonio.

bioactive. The occurrence of carotenoids within peppers is diverse, with the identification of about 50 distinct compounds within different *Capsicum* species and cultivars.[1] Although most of these carotenoids, such as β-carotene, commonly occur in other vegetables, there are two xanthophylls that are biosynthesised exclusively in peppers. These are capsanthin and capsorubin, both belonging to the red fraction (Figure 4.1).[6]

There are a variety of colours in peppers, from green to yellow, orange, red and even brown (Figure 4.2). Such variations are caused by distinct combinations of the concentration and qualitative chemical profiling of carotenoids. The carotenoid content within *Capsicum* is diverse and highly influenced by ripening and other factors, which we will explore later in this chapter. In qualitative terms, each colour has a major set of compounds that are distinct from one another. The major carotenoids of the green and yellow cultivars are violaxanthin, antheraxanthin and lutein, whereas red cultivars present capsanthin, capsorubin and cryptoxanthin as their major carotenoids. The concentrations of the major carotenoids are inversely proportional in yellow and red peppers, mainly due to the capsanthin and capsorubin produced by enzymatic oxidative reactions involving antheraxanthin and violaxanthin, respectively.[7–9]

Figure 4.2 The distinctive colours of *Capsicum* fruits. Image credit: Carlos Magno da Silva Antonio.

In general, red-coloured peppers have a higher concentration of carotenoids than yellow peppers. In addition to the xanthophylls, reddish fruits also have a greater diversity of carotenoids and higher concentrations of β-carotene. Yellow peppers are unable to produce red fraction carotenoids, giving them a lower chemodiversity. Although α- and β-carotenes are classified as yellow fraction carotenoids, they are responsible for the orange colour within fruits and are the major carotenoids in cultivars such as habanero fatalii (Figure 4.3) and in highly pungent cultivars such as Trinidad scorpion. In fact, red and orange fruits present a higher concentration of β-carotene than their equivalent yellow fruit cultivars.[10] The concentration of β-carotene in chilli peppers is of great interest. This metabolite has a high nutritional value, not only for its antioxidative properties, but because the human body uses it as a precursor of vitamin A. The accumulation of β-carotene within peppers can reach concentrations up to 1310.0 mg per 100 g fruit.[7,9]

4.2.2 Modification of the Carotenoid Content of Chilli Peppers

The carotenoid profile of chilli peppers is highly modified by several endogenous and exogenous factors, one of which particularly stands out: the ripening process. During ripening, the carotenoids in chilli peppers are esterified with fatty acids. Immature fruits therefore have a greater content of free carotenoids, whereas mature fruits have a greater concentration of esterified carotenoids.[11]

The esterification process can increase the chemical and thermal stabilities of carotenoids, enabling their accumulation in fruits. A high content of carotenoids during maturation is desirable to the fruit because they act as photoprotectors, precursors of other essential compounds and in the reproductive mechanisms of the plant, attracting small animals to promote seed dispersal and pollination. For instance, mature peppers can have a total carotenoid content 10–90 times higher than immature peppers. Carotenoid esterification mainly occurs with short chain fatty acids, such as lauric, myristic, palmitic and linoleic acids.[9]

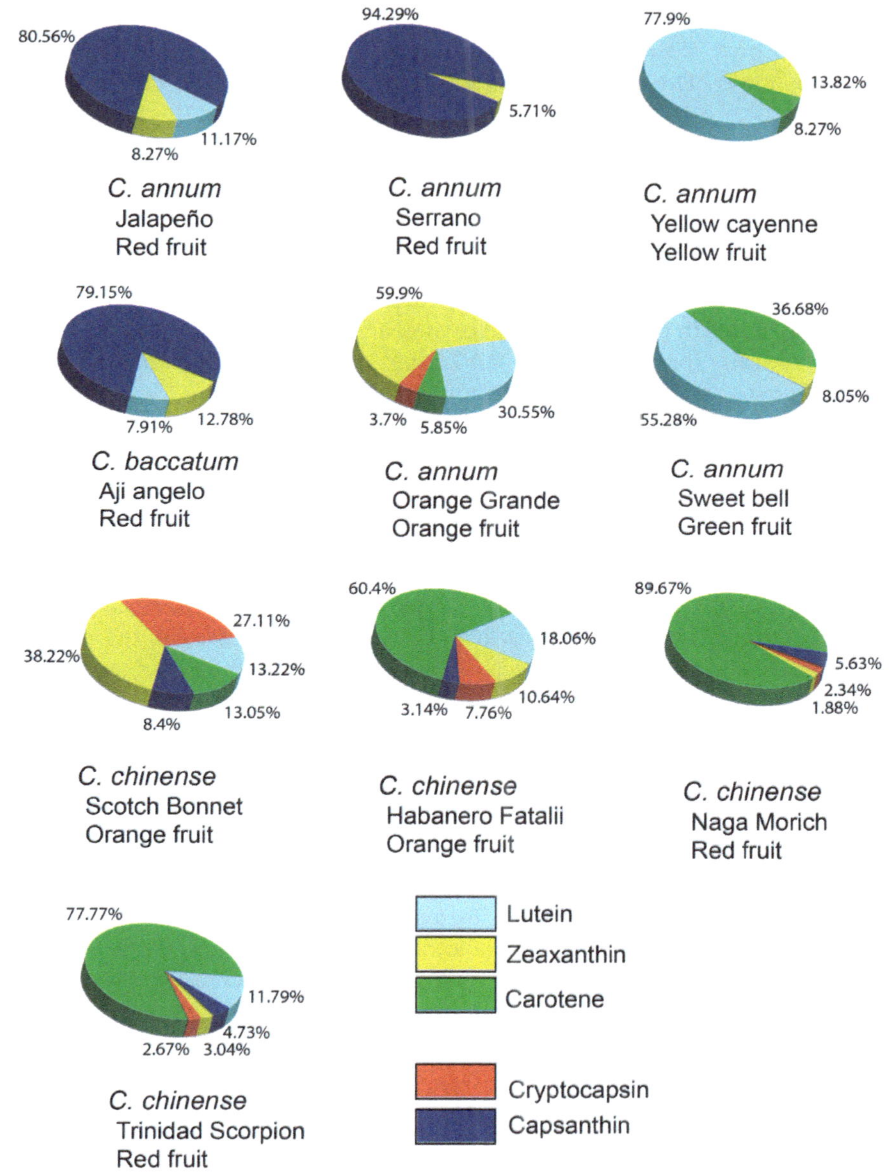

Figure 4.3 Chemical structures of the yellow and red fractions of carotenoids in different coloured peppers. Image credit: Carlos Magno da Silva Antonio.

In general, unripe peppers mainly contain yellow fraction carotenoids, which are biosynthesised in the chloroplast and are involved in photosynthesis. As ripening proceeds, the biosynthesis of chloroplast pigments is partially or totally interrupted because the ripe fruit does not require photosynthesis. In the case of yellow-coloured fruits, the major compounds in the carotenoid profile when full maturity is achieved are lutein and β-carotene or violaxanthin and α-carotene. In both cases, the carotene content is lower. The concentration of lutein and violaxanthin in ripe yellow fruits can be up to 67% of the total carotenoid content. In red-coloured fruits, the yellow fraction carotenoids are substituted by capsanthin, capsorubin, and cryptocapsin during ripening, creating more variation in the carotenoid profile during the life cycle of the fruit.[1,7]

Similar to capsaicinoids, the carotenoid content and profile also vary due to environmental factors and understanding how this occurs is fundamental to the development of cultivation methods for specific uses of *Capsicum*. As the chemicals in this class act as thermal and photoprotectors in the plant, biomes with higher temperatures and a higher incidence of radiation result in an increase in the total carotenoid content, particularly xanthophylls. Raising the temperature is more effective in increasing the total carotenoid content during cultivation than increasing the amount of solar radiation.[7,9]

In addition to their antioxidative properties, xanthophylls are also the precursor of abscisic acid, a hormone that regulates the closure of stomates. The stomates are pores responsible for the exchange of gases between a plant and the surrounding atmosphere. Plants release water vapour through stomates and absorb carbon dioxide. A dry environment causes the plant to close its stomates to avoid water loss. Under these conditions, the xanthophyll concentration increases because it is required by the plant to regulate these structures.[1]

4.2.3 Apocarotenoids

A chemical class closely related to carotenoids, yet poorly studied in *Capsicum*, is the apocarotenoids.[12] The apocarotenoids are a carotenoid-derived class of chemicals produced by the oxidative cleavage of carotenoids that contain <40 carbon atoms. The formation of apocarotenoids *in natura* involves enzymes known as carotenoid cleavage oxygenases, such as the carotenoid cleavage dioxygenases and 9-*cis*-epoxycarotenoid dioxygenases. Depending on the enzyme, a distinct apocarotenoid will be formed, which can be non-volatile or volatile.[13] The more frequently detected apocarotenoids within *Capsicum* are derivates of zeaxanthin, carotene and the red fraction of carotenoids (Figure 4.4). The increasing interest in apocarotenoids is related to their health-promoting functions, biological role and effects on the colour and taste of *Capsicum* fruit.

The antioxidative functions of the apocarotenoids commonly detected in saffron and *Citrus* species are under investigation as possible chemopreventive agents.[14] In plants, apocarotenoids are related to growth regulation, the formation of branches, seed dispersion and self-pollination. Apocarotenoids are also known to influence the taste and aroma of vegetables. However, their increased accumulation in *Capsicum* bioproducts is also related to a loss of colour and quality because formation above natural concentrations represents the degradation of carotenoids.[12,15] It was observed that drying at 35 °C for

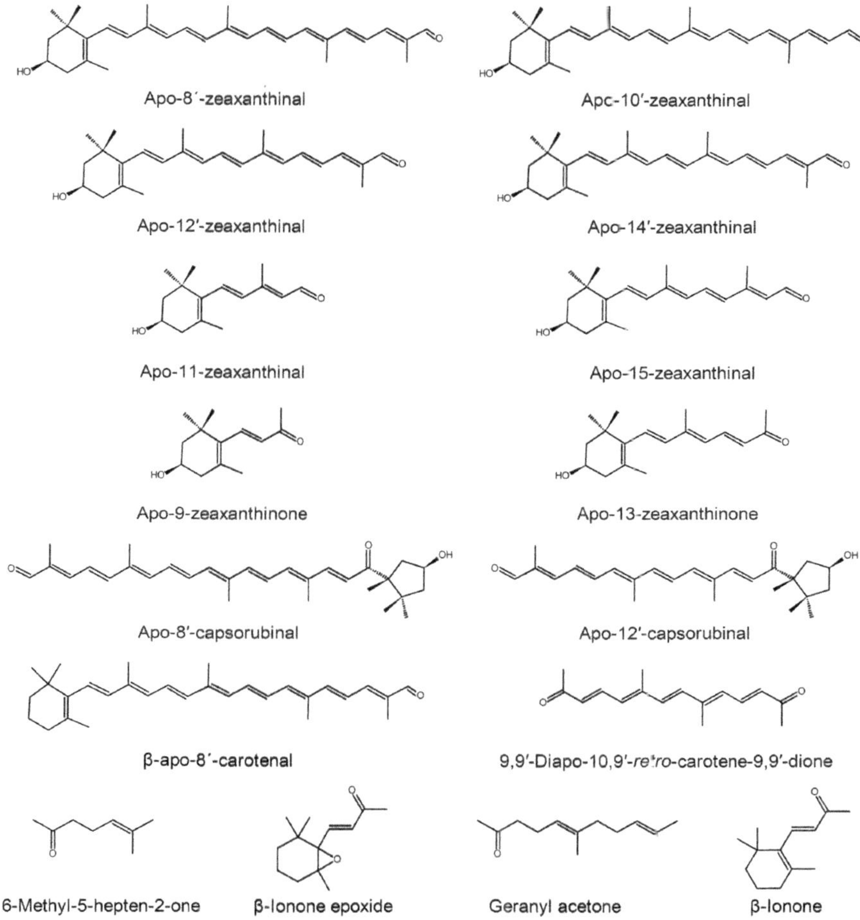

Figure 4.4 Apocarotenoids detected in *Capsicum*. Image credit: Carlos Magno da Silva Antonio.

three weeks significantly increased the apocarotenoid content, along with a loss of red colour perception in fruit – for instance, the β-ionone epoxide in one of the phenotypes studied increased its concentration from 319.38 to 2579.14 ng mL^{-1} during drying.[15]

Most of the data available for *Capsicum* apocarotenoids is qualitative and only few varieties of *Capsicum annuum*, *Capsicum chinense* and *Capsicum baccatum* have been evaluated.[12,16,17] The absence of quantitative studies is a consequence of their minor concentration in the genus (Table 4.1). A study of 15 varieties of *Capsicum* clearly showed the different concentrations of carotenoids and apocarotenoids: the maximum concentration of β-citraurin was 0.610 μg g^{-1}, whereas the maximum concentrations of β-carotene and capsanthin were 261.579 and 66.993 μg g^{-1}, respectively.[12] The quantitative

Table 4.1 Concentration of selected apocarotenoids in *Capsicum*.

Apocarotenoid	Concentration		Sample	Ref.
6-Methyl-5-hepten-2-one	47.22	ng mL^{-1} fresh fruit	Unidentified cultivar R2	15
	25.02		Unidentified cultivar R4	
β-Ionone	11.85		Unidentified cultivar R2	
	1.74		Unidentified cultivar R4	
β-Ionone epoxide	319.38		Unidentified cultivar R2	
	44.33		Unidentified cultivar R4	
Geranyl acetone	13.73		Unidentified cultivar R2	
	4.04		Unidentified cultivar R4	
β-Citraurin (apo-8′-zeaxanthinal)	0.092	μg g^{-1} dry fruit	*C. baccatum* var. ají limòn	12
	0.61		*C. chinense* var. habanero red savina	
	0.142		*C. chinense* var. naga yellow	
	0.09		*C. annuum* var. jalapeño	
β-Apo-8′-carotenal	0.052		*C. baccatum* var. ají limòn	
	0.336		*C. chinense* var. habanero red savina	
	0.05		*C. chinense* var. naga yellow	
	0.123		*C. annuum* var. jalapeño	

and qualitative profiles of apocarotenoids in *Capsicum* are also influenced by the species and variety, similar to carotenoids, although only Zoccali *et al.*[12] have evaluated the modification of the apocarotenoid profile within distinct varieties.

4.3 Vitamins

Capsicum fruits are consumed in many different diets and cultures. Their importance is due to their incredible nutritional profile. *Capsicum* species are a rich source of several chemicals that are essential to the normal biological functioning of mammals, but are not biosynthesised by them. The chemical structures of vitamins are not homogeneous, although they can be classified in two groups: fat-soluble and water-soluble vitamins.

Capsicum peppers are recognised as a reliable source of vitamins A, E, C and folate.[18] Vitamin A is a fat-soluble vitamin also known as retinol (Figure 4.5). There are several analogues of vitamin A, which are classified as retinoids. Retinol is usually emphasised because it is the bioactive form present in the tissues of mammals. Another retinoid is its aldehyde analogue, retinal. Retinoids are composed of four isoprene units and are derived from carotenoids, such as carotene; the carotenoids α/β-carotene and β-cryptoxanthin are both classified as provitamin A carotenoids (the precursors of retinoids). The retinol content of peppers is often reported indirectly as the concentration of these carotenoids, either individually or associated, as retinol-equivalent units.[19–22] Because retinol is produced by provitamin A carotenoids, variations in the vitamin A content of *Capsicum* peppers are influenced by the same factors that affect the carotenoid content. The concentration of retinol itself varies widely in *Capsicum* from

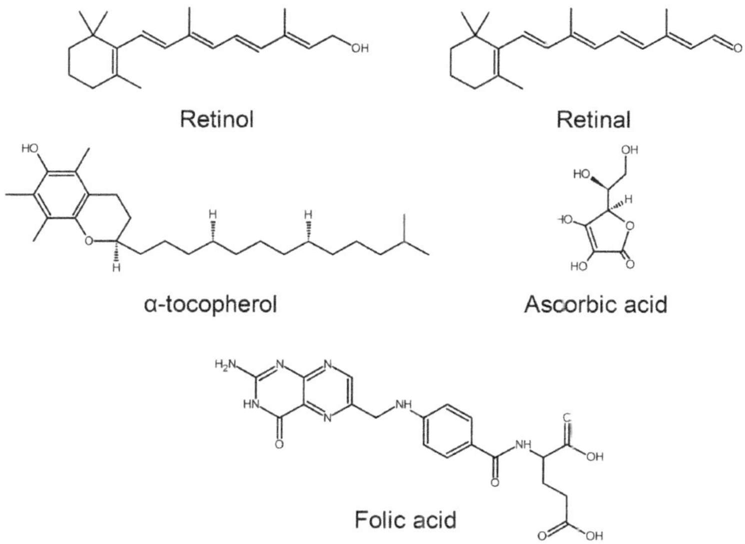

Figure 4.5 Major vitamins in *Capsicum* peppers. Image credit: Carlos Magno da Silva Antonio.

10 to 6250 μg 100 g^{-1} depending on the species, variety or cultivar.[18,23] Provitamin A as retinol-equivalent units has been reported in the range 18–402 μg 100 g^{-1}, although lower concentrations are usually observed in low-pungency or immature fruits.[19,20]

Vitamin E is a fat-soluble vitamin, mainly represented by α-tocopherol (Figure 4.5). Vitamin E and tocopherols are the name of a group containing eight compounds derived from 6-chromanol. Their importance is mainly related to their antioxidative activity, which is related to the prevention of cell damage and cancer.

The concentration of tocopherol varies with the maturity of the fruit, the environmental conditions, the species and the cultivation/harvesting methods. It is one of the chemical features evaluated in breeding programmes to produce more nutritional cultivars.[24,25] The tocopherol content increases with maturity and has a positive correlation with the carotenoid content.[24]

The total vitamin E content of *Capsicum* peppers is similar to that in other common vegetables and fruits, such as potatoes, sunflowers, carrots and apples,[26] although each variety has a distinct content (Table 4.2). For instance, *C. baccatum* and *Capsicum pubescens* are domesticated species with a lower total tocopherol content.[25,27] The different types of tocopherol present different concentrations in specific organs. In general, the highest concentrations of γ-tocopherol are in the seeds, whereas the highest concentrations of α-tocopherol are in the pericarp.[25,28] α-Tocopherol is the major form of vitamin E in *Capsicum*, with a concentration ranging from 0.23 to 45 mg 100 g^{-1}.[25,26,29]

Table 4.2 Tocopherol concentration in *Capsicum* fruits (mg 100 g^{-1} dry fruit).

Species	Variety or cultivar	Tocopherol α	β	γ	Ref.
C. annuum	AC 1979	10	4.3	—	29
C. annuum	Cayenne	6	1.9	—	
C. annuum	Bell pepper	13.81	—	—	26
C. annuum	Fire flame	31.683	—	8.6	28
C. annuum	Serrano	—	1.31	3.167	24
C. baccatum	Ají limòn	—	0.242	0.17	
C. chinense	AC2212	16.32	1.6	—	29
C. frutescens	Korba	40	—	<10	30
C. pubescens	PER007276	5.9	—	1	27
C. pubescens	PER017948	16.7	0.1	0.1	
Mean total content[27]					
C. pubescens	10				
C. annuum	25				
C. baccatum	17				
C. chinense	11				
C. frutescens	11				

Vitamin C is a water-soluble vitamin with the chemical name ascorbic acid (Figure 4.5). It is a smaller molecule than either vitamin A or E and is involved in several biological processes, including electron transport reactions responsible for the generation of energy. High concentrations of vitamin C are usually associated with *Citrus* species (*e.g.* oranges and lemons) and few people are aware that *Capsicum* peppers have a similar or even higher vitamin C content.

For instance, kiwi (*Actinidia deliciosa*) is considered to be one of the fruits with the highest vitamin C content with, on average, 93 mg ascorbic acid per 100 g of fruit. Oranges have an average concentration of 53 mg 100 g^{-1}.[31] The vitamin C content of *Capsicum* can reach an incredible 195 mg 100 g^{-1}.[18] It is not common to see people suggesting the ingestion of chilli peppers as a source of vitamin C because concentrations >100 mg of ascorbic acid per 100 g are found in the highly pungent varieties, such as *C. chinense* Trinidad 7 pot, naga morich and bhut jolokia. Less pungent varieties, such as the Spanish cayenne, ají limòn, bell pepper and red roccoto have vitamin C contents closer to oranges, ranging from 24 to 70 mg 100 g^{-1}.[18]

The vitamin C content of chilli peppers is correlated with the colour and stage of maturity. In general, green varieties and immature fruits have the lowest concentrations.[10] The maturity stage implies that the plant will have had a longer exposure to sunlight and changes in the amount of solar radiation over several seasons. Because ascorbic acid is a photoprotector used by plants, its concentration will be higher during ripening and under conditions of oxidative stress.[3,22]

Folic acid is a water-soluble vitamin also known as vitamin B9 and occurs in plants as folate (Figure 4.5). Vitamin B9 is involved in the biosynthesis of amino acids and in the activation of tRNA. The folate concentration in *Capsicum* varies widely within genotypes, with an average level of 70 μg 100 g^{-1} fruit. Some sweet bell pepper cultivars have concentrations as low as 8 μg 100 g^{-1}.[32,33] Some *C. annuum* cultivars, such as shishito, pasilla and Amish chicken, have a higher folate content than spinach (193 μg 100 g^{-1}).[18,34] The folic acid content is highly influenced by post-harvesting processes, such as the drying and preservation methods.[34]

4.4 Fatty Acids

Fat-soluble vitamins are transported from plant organs *via* lipids, which are hydrophobic molecules composed of carbon, hydrogen and oxygen atoms. Among the several different types of lipid present in plants, fatty acids are particularly interesting. A number of biological functions in mammals depend on essential fatty acids. One of the most important of the essential fatty acids is linoleic acid, which is not biosynthesised by mammals, but which has an essential role in energy generation and metabolism. Fatty acids are aliphatic carbon chains that can be either linear or branched, saturated or unsaturated (Figure 4.6). In plants, fatty acids mainly occur as unsaturated carbon chains with one or more carbon–carbon double bonds.

Branched chain fatty acids are part of the biosynthetic pathway of capsaicinoids. Their presence in *Capsicum* not only provides nutrition, but also helps to make the genus unique. Unsaturated fatty acids represent >80% of the total lipid content of *Capsicum*,[35–38] with average linoleic and oleic acid concentrations of 70 and 10% w/w, respectively (Table 4.3).[37] The linoleic acid concentration varies from 300 to 600 mg 100 g^{-1} fruit. Palmitic acid is the major fatty acid in the saturated fraction.[35,36]

4.5 Volatile Organic Compounds

As their name suggests, VOCs are organic chemical compounds produced in the gas phase or which evaporate quickly at normal temperatures and pressures (20 °C/293.15 K/68 °F and 1 atm/14.696 psi/101.325 kPa). These compounds are carbon-based structures with a molecular mass usually <250 Da. The VOCs include a wide range of chemicals, including isoprenoids, sesquiterpenes, ketones, aldehydes, alcohols, methyl ethyl ketones, acetaldehydes, nitrogen-containing compounds and furanocoumarins. In plants, VOCs are not only responsible for aroma, but are also involved in defence, thermal resistance and reproduction.

The aroma of peppers is a complex mixture of VOCs (Figure 4.7). Most of the characterisation of *Capsicum* VOCs is based on *C. annuum*, the major commercial species in the genus. Even with the relatively low scientific exploitation of such specialised metabolites, 136 distinct VOCs have been detected in *Capsicum* fruits.[41,42] VOCs can be observed both in fresh fruit and in dry powders, although, depending on the conditions, the VOCs of dry powders (*e.g.* paprika) significantly decrease in concentration after three

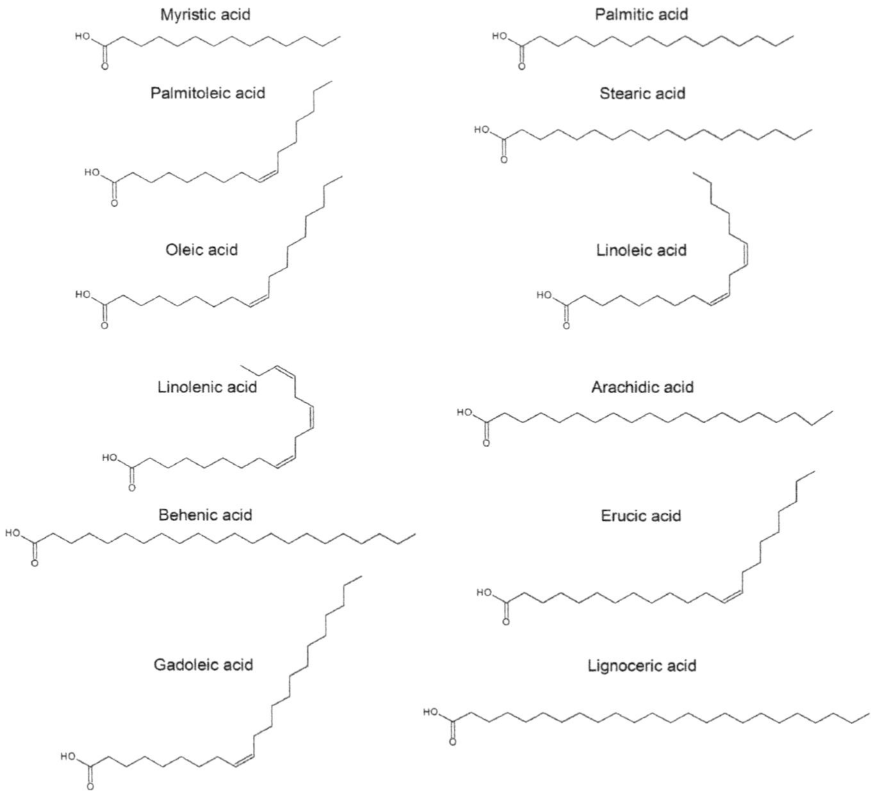

Figure 4.6 Fatty acids detected in *Capsicum* species. Image credit: Carlos Magno da Silva Antonio.

months of storage or during aggressive drying methods.[42,43] The total VOC concentration in peppers can range from 6.041 to 113.16 mg kg^{-1} dry fruit weight, considerably lower than the accumulation of only one of the carotenoids (β-carotene) produced by *Capsicum* fruits. Despite this, VOCs cannot be underestimated because they are essential in plant development, have bioactive properties and smell pleasant.

The most important VOCs in peppers are the terpenes and pyrazines.[41,44] The VOC content of peppers is influenced by several biotic and abiotic factors. For instance, their concentration in peppers is higher in pungent and red or orange species because most VOCs, such as isoprenoids, aldehydes and esters, are involved in the biosynthetic pathways that produce carotenoids and capsaicinoids. In addition, the total VOC content increases during ripening, with significant changes in their qualitative profile. For instance, the major VOCs in hot ripe fruits are terpenoids, aldehydes and esters, such as 1,4-cineole, geraniol, cubenol, cadinol, himachalol and geranyllinalool.[41,44,45]

Table 4.3 Content of major fatty acids in *Capsicum* fruits.

Species	Variety	Linoleic acid	Oleic acid	Palmitic acid	Units	Ref.
C. annuum	Cayenne	270.26	98.13	202.93	mg 100 g^{-1} of pulp	37
C. annuum	Bell pepper	292.75	158.12	174.53		
C. baccatum	Dedo de moça	302.8	161.69	173.4		
C. chinense	Red habanero	209.03	72.06	151.52		
C. chinense	Orange habanero	272.96	26.51	188.08		
C. frutescens	Malagueta	352.35	232.44	200.49		
C. annuum	Glabriusculum	6.09	75.88	14.43	% w/w of seed	39
C. pubescens		7.58	76	10.62		
C. chinense	Naga king	60.43	8.22	15.66	g 100 g^{-1} of total fat in whole fruit	40

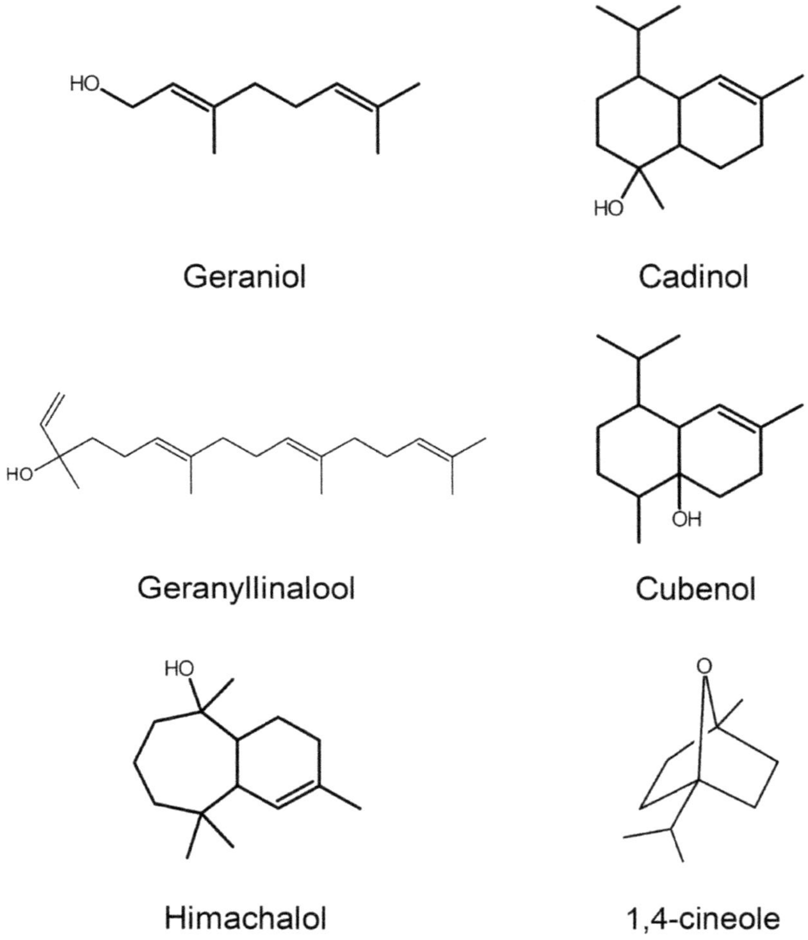

Figure 4.7 Chemical structures of some of the VOCs present in *Capsicum* fruits. Image credit: Carlos Magno da Silva Antonio.

Each species and cultivar has a distinct VOC profile. The major characteristic VOCs in commercial cultivars are given in Table 4.4. The distinct VOC profiles of pungent species can be used to distinguish them – for example, *C. baccatum*, *C. chinense*, *C. pubescens* and *C. annuum* cultivars have been fully distinguished based solely on their composition of volatile metabolites.[46] Although VOCs are easily lost by evaporation from natural peppers and their bioproducts, their presence in derived natural products is used as a quality factor. The retention of VOCs in pepper products indicates that the peppers were not exposed to thermal stress during processing and were correctly stored, preserving their biological properties. An example of the use of VOCs in quality control during harvesting is the terpene content. The use of pesticides can result in a decrease in the terpene concentration in the fruit of chilli peppers because the plant produces more terpenes as

Table 4.4 Volatile organic chemical markers of *Capsicum* varieties.

C. annuum L. cv. Inan3363	*C. chinense* habanero	*C. annuum* jalapeño	*C. chinense* × *C. frutescens* bhut jolokia	*C. pubescens* rocoto	*C. baccatum* ají amarillo
Limonene	6-Methyl-4-heptenyl 3-methyl butanoate	2-Hexenal	4-Methyl pentyl 3-methyl butanoate	2-Hexenoic acid	γ-Terpinene
2-Methylpropanol	6-Methyl-4-heptenyl 2-methyl propanoate				
Ethanol	3,2-Methyl butanoate				
3-Methylbutanol	6-Methyl-4-heptenol				
3-Hydroxy-2-butanone	2-Hexenal				
N-Butyl-isobutyrate	3-Methylbutyl 8-methyl-6-nonenoate				
Acetic acid	β-Farnesene				
Isododecane	Octyl 3-methyl butanoate				
Styrene	3-Methyl butanoic acid				
2-Furanmethanol					
2,6-Dimethylpyrazine					
Dimethyl sulfide					
Pyranone					
Isoeugenol					

protection against biotic and abiotic stressors.[44] The VOCs in *Capsicum* have a pleasant smell and terpenes and pyrazines are bioactive, with anticancer, antioxidant, antiviral, anti-inflammatory, antiparasitic, antifungal and antimicrobial activity.[47–49]

4.6 Phenolic Compounds

Another major class of specialised metabolites produced by peppers is the phenolic compounds. These metabolites are widely produced in plants because they can help them to withstand biotic and abiotic stressors – for example, phenolic compounds can reduce oxidative stress. These compounds, also known as polyphenols, have at least one aromatic ring and one hydroxyl group. Examples of natural phenolic compounds include small molecules such as gallic acid and ferulic acid, which are the main precursors of important biosynthetic routes to more complex structures, such as condensed tannins.[7]

The most prominent phenolic compounds in chilli peppers are the vanillic, caffeic and ferulic acids (Figure 4.8) with only one aromatic ring, and the flavonoids, which have at least three aromatic rings. Despite the great importance of phenolic acids, flavonoids have higher concentrations in peppers and are more prominent in their chemical profile.[7,50]

4.6.1 Flavonoids

Flavonoids as a class consist of >7000 distinct substances derived from the phenylpropanoid pathway. Their chemical structure is based on two aromatic rings and one heterocyclic ring containing an oxygen atom. As with carotenoids, the double bond system across the flavonoid structure enables interactions with visible light, conferring chromogenic properties. Because of this, flavonoids are also responsible for the coloration of peppers. Flavonoids also aid plant growth, seed dispersion, defence against predators and protection against ultraviolet radiation, temperature and drought stress.[1,50] Flavonoids have a high diversity of structures, with distinct numbers of hydroxyl moieties, methoxyl moieties and double bonds and opening of the heterocyclic ring. Flavonoids can be classified as flavones, flavonols, isoflavones, anthocyanins and chalcones.[50]

Figure 4.8 Major phenolic acids present in *Capsicum* species. Image credit: Carlos Magno da Silva Antonio.

The evaluation of flavonoids in chilli peppers is important because they have nutritional functions in humans and can only be obtained through the diet. Several pharmaceutical studies have emphasised that flavonoids can be used as anticholinesterase, anti-inflammatory, antimicrobial and anticarcinogenic agents.[51,52] Flavonoids also have roles as antioxidants and photoprotectors in plants and are involved in photosynthesis and, as a result, they mainly accumulate in the peels and leaves of peppers. The action of flavonoids against oxidative stress means that their qualitative and quantitative profiles in peppers are susceptible to environmental conditions, such as the incidence of ultraviolet radiation and soil fertility.[7,53]

The major flavonoids within chilli peppers are quercetin and luteolin derivates (Figure 4.9), mainly the *O*-glycosyl types. The *O*-glycosyl flavones are the major type of flavonoids in most *Capsicum* fruits during different stages of ripening and cultivation methods. In addition, they are sensitive environmental factors and are often neglected in the scientific exploitation of chilli peppers.[22,50,54]

The total flavonoid content can range from 0.0036 to 0.2024 mg g^{-1} fresh fruit, with the quercetin derivates representing a large part of this content (Table 4.5).[50] The flavonoid profile within *Capsicum* varies as a result of biotic factors, such as the cultivar, variety and stage of ripening; the total flavonoid content varies within genotypes. As a result of their nutraceutical importance, the flavonoid content is usually studied in low-pungency cultivars. The sweet genotypes of *C. annuum*, *C. baccatum* and *C. frutescens* generally contain more flavonoids than the highly spicy *C. chinense* varieties. The flavonoid content of sweet genotypes can be up to ten times higher than that of the hot genotypes.[54,55]

Flavonoids are highly influenced by ripening. For instance, it is estimated that the total flavonoid content can decrease by up to 85% during ripening because the mature stage does not require intense photosynthetic activity and there is an accumulation of other antioxidant metabolites, such as carotenoids and capsaicinoids.[50,54] It is a little more difficult to predict the dependency of flavonoids on the stage of ripening.

In red fruit cultivars, the content of flavonoids increases until the fruit reaches the maturity stage.[54,56] This maturity stage is characterised by the initial loss of the green colour. In the transition stage, the fruit can have two colours (green and red) or even be orange. As the fruit reaches full maturity, with a homogenous red colour, the flavonoid content decreases by up to five times its concentration in the immature stages.[55,57–59] Modification of the flavonoid content of *Capsicum* fruit can be attributed to several processes, each including the production of other phenolic compounds, enzymatic degradation and the competitive use of precursors to the biosynthesis of capsaicinoids. In sweet peppers, the flavonoid content is more influenced by the variety. Although *C. annuum* cultivars, such as jalapeño, have an increased flavonoid content in the mature stages,[54] other varieties, such as versaga, have a decreased content with maturation.[22]

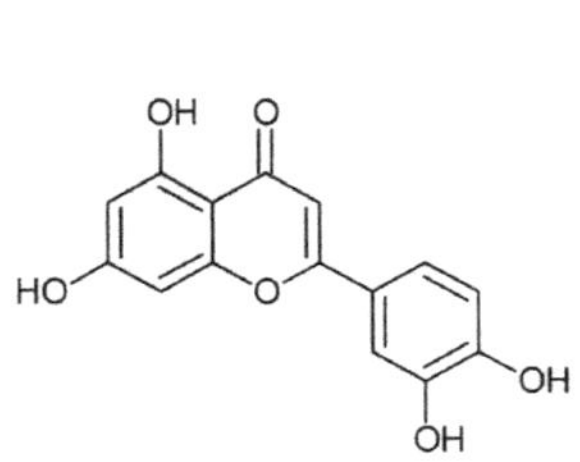

Luteolin

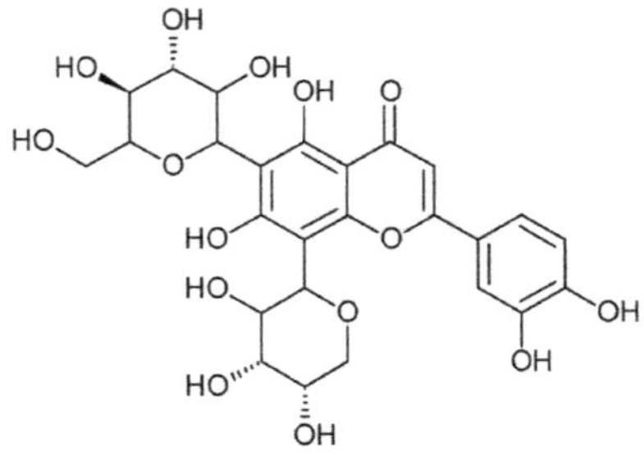

Luteolin-*C*-*β*-*D*-glucopyranoside -8-*C*-α-L-arabinopyranoside

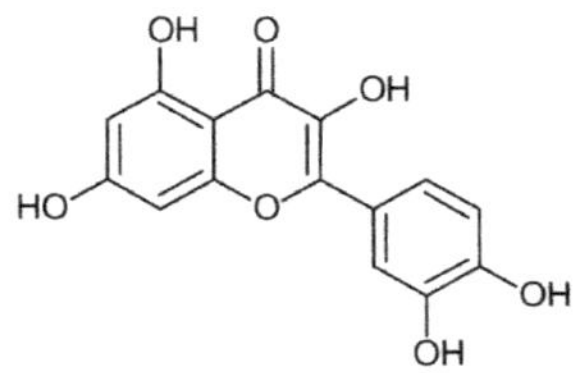

Quercetin

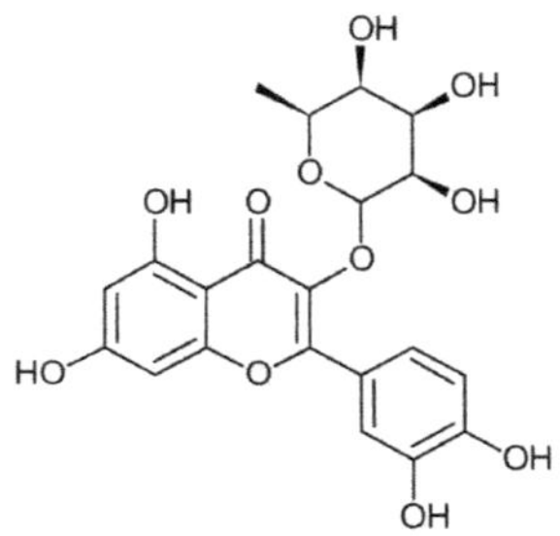

Quercetin-3-*O*-*α*-*L*-rhamnopyranoside

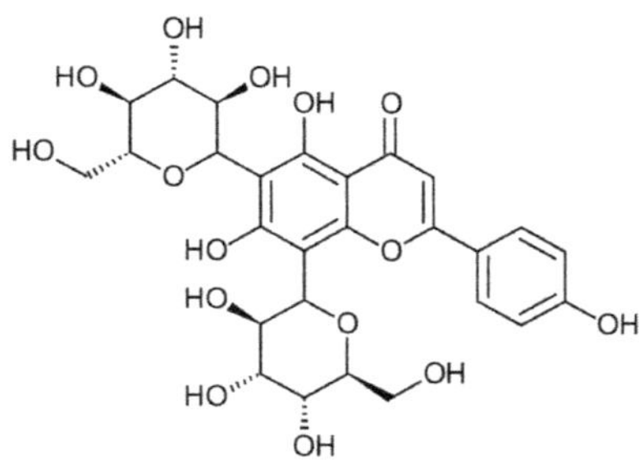

Apigenin-6-C-glucoside 8-C-arabinoside

Figure 4.9 Major flavonoids in *Capsicum* species. Image credit: Carlos Magno da Silva Antonio.

Table 4.5 Major flavonoids in the chemical profile of *Capsicum annuum*.

Compound	Concentration ($mg\,kg^{-1}$ of fruit)[7]
Quercetin glucoside	19 860
Quercetin 3-*O*-α-L-rhamnoside-7-*O*-β-D-glucoside	12 300
Quercetin	10 810
Kaempferol pentosyldihexoside	17 170
Luteolin	880
Luteolin 7-*O*-[2-(β-D-apiosyl)-β-D-glucoside]	5660
Luteolin-6-*C*-glucoside	5090
Apigenin 6-*C*-β-D-glucoside-8-*C*-α-L-arabinoside	900

4.7 Reducing Sugars

The taste of some *Capsicum* species and their nutritional value is related to their chemical composition. The sweet taste of some cultivars is related to high content of natural sugars. Naturally occurring sugars are not just tasty, but also have antioxidant and antimutagenic properties. This is the main reason for the interest in natural foods that have a high sugar content, especially if it consists of reducing sugars.

From a chemical perspective, reducing sugars are carbohydrates that contain a free aldehyde or ketone group in their chemical structure. The presence of these functional groups in the molecule enables it to act as a reducing agent in oxidation–reduction reactions – that is, these sugars can easily donate an electron to an unstable molecule to reduce its reactivity. This reduces oxidative stress in living organisms, a metabolomic disorder responsible for several diseases. Examples of reducing sugars include glucose, fructose, lactose and maltose. The major reducing sugars studied in *Capsicum* species are glucose and fructose (Figure 4.10).

Reducing sugars are usually presented as their total content (Table 4.6).[60–63] The content of reducing sugars is closely related to both biotic and abiotic factors. As a result of their nutritional importance, tracking cultivars with high contents of reducing sugars is desirable in breeding programmes. Most of the flavonoids in *Capsicum* are glycosylated derivates. The content of reducing sugars is strongly positively correlated with the total phenolic content and increases during the ripening process.[64]

4.8 Concluding Remarks

Capsicum peppers are one of the major spices used by humans. Their unique taste, aroma, nutritional and medicinal properties have drawn attention to the use of chilli peppers in various forms. Peppers are usually only discussed and highlighted because of their pungency, but their other remarkable properties are not only a result of their capsaicinoid content. A complex chemical profile is behind each distinct colour, aroma and bioactivity. Their pharmacological interest is deeply related to their carotenoid, flavonoid and VOC (*e.g.* sesquiterpene) content. We need to understand the complex

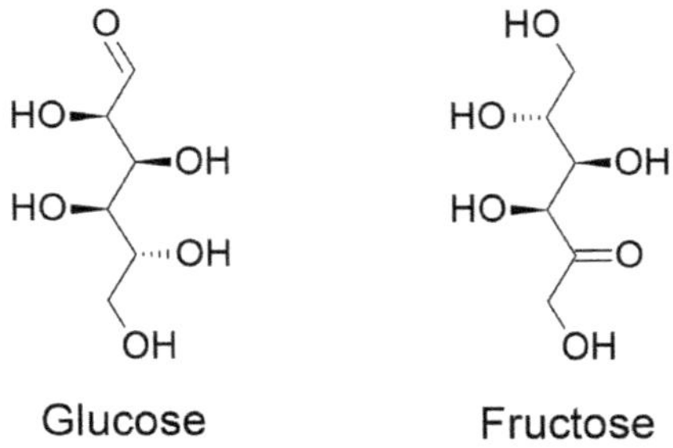

Figure 4.10 Chemical structures of glucose and fructose. Image credit: Carlos Magno da Silva Antonio.

Table 4.6 Total reducing sugar content in *Capsicum* species.

Species	Variety/cultivar	Total reducing sugars	Units	Ref.
C. annuum	Cheonyang	0.692	$mg\,g^{-1}$ fresh weight	60
C. annuum		15.82	$mg\,g^{-1}$ fresh weight	64
C. baccatum		11.64	$mg\,g^{-1}$ fresh weight	64
C. baccatum var. pendulum	Cambuci	5.9	% w/w fruit	65
C. chinense		14.59	$mg\,g^{-1}$ fresh weight	64
C. frutescens		8.68	$mg\,g^{-1}$ fresh weight	64
C. frutescens	Malagueta	4.91	% w/w fruit	66
C. frutescens	Average of 40 genotypes	1.55	% w/w fruit	61

chemical networking within a plant to fully comprehend and preserve its nature while taking advantage of it for our survival.

The biodiversity within the *Capsicum* genus produces a wide chemodiversity. Each species, cultivar and variety has a distinct profile of specialised metabolites. Each growing, harvesting and post-harvesting method generates a distinct profile of flavonoids and carotenoids. Abiotic stressors and the ripening process affect the chemical profile of peppers. *Capsicum* fruits can therefore be cultivated and processed in a way to enhance the properties desirable for different purposes.

References

1. A. S. Antonio, L. S. M. Wiedemann and V. F. Veiga Junior, *RSC Adv.*, 2018, **8**, 25767–25784.
2. N. Riquelme and S. Matiacevich, *CyTA-J. Food*, 2017, **15**, 344–351.
3. H. Bae, G. K. Jayaprakasha, K. Crosby, K. S. Yoo, D. I. Leskovar, J. Jifon and B. S. Patil, *J. Food Compos. Anal.*, 2014, **33**, 195–202.
4. G. Zakynthinos and T. Varzakas, *Curr. Res. Nutr. Food Sci.*, 2016, **4**, 38–51.
5. D. B. R. Amaya, *Food Carotenoids: Chemistry, Biology and Technology*, IFT Press, John Wiley & Sons, Ltd, New Jersey, 2015.

6. I. Guzman, S. Hamby, J. Romero, P. W. Bosland and M. A. O'Connell, *Plant Sci.*, 2010, **179**, 49–59.
7. N. de Sá Mendes and É. C. Branco de Andrade Gonçalves, *Trends Food Sci. Technol.*, 2020, **99**, 229–243.
8. T. S. Agostini-Costa, I. S. Gomes, L. A. M. P. de Melo, F. J. B. Reifschneider and C. S. da Costa Ribeiro, *J. Food Compos. Anal.*, 2017, **57**, 73–79.
9. N. M. Hassan, N. A. Yusof, A. F. Yahaya, N. N. M. Rozali and R. Othman, *Antioxidants*, 2019, **8**, 1–25.
10. F. Fratianni, A. D'acierno, A. Cozzolino, P. Spigno, R. Riccardi, F. Raimo, C. Pane, M. Zaccardelli, V. T. Lombardo, M. Tucci, S. Grillo, R. Coppola and F. Nazzaro, *Antioxidants*, 2020, **9**, 1–16.
11. D. Penagos-Calvete, J. Guauque-Medina, M. F. Villegas-Torres and G. Montoya, *Hortic., Environ. Biotechnol.*, 2019, **60**, 227–238.
12. M. Zoccali, D. Giuffrida, F. Salafia, F. Rigano, P. Dugo, M. Casale and L. Mondello, *Food Chem.*, 2021, **334**, 127595.
13. J. C. M. Beltran and C. Stange, in *Carotenoids in nature: biosynthesis, regulation, and function*, ed. C. Stange, Springer International Publishing, Cham, 2016, vol. 79, pp. 377–414.
14. E. H. Harrison and L. Quadro, *Annu. Rev. Nutr.*, 2018, **38**, 153–172.
15. H. M. Berry, F. Lai, A. Kende, D. V. Rickett, C. J. Baxter, E. M. A. Enfissi and P. D. Fraser, *Food Chem. Mol. Sci.*, 2021, **2**, 100013.
16. T. Maoka, Y. Fujiwara, K. Hashimoto and N. Akimoto, *J. Agric. Food Chem.*, 2001, **49**, 3965–3968.
17. D. Giuffrida, M. Zoccali, S. V. Giofrè, P. Dugo and L. Mondello, *Food Chem.*, 2017, **231**, 316–323.
18. M. B. Kantar, J. E. Anderson, S. A. Lucht, K. Mercer, V. Bernau, K. A. Case, N. C. Le, M. K. Frederiksen, H. C. DeKeyser, Z. Z. Wong, J. C. Hastings and D. J. Baumler, *PLoS One*, 2016, **11**, 1–12.
19. E. Rodríguez-Rodríguez, M. Sánchez-Prieto and B. Olmedilla-Alonso, *Food Chem. X*, 2020, **6**, 100092.
20. V. M. Russo and L. R. Howard, *J. Sci. Food Agric.*, 2002, **82**, 615–624.
21. T. L. Olatunji and A. J. Afolayan, *Food Sci. Nutr.*, 2018, **6**, 2239–2251.
22. A. Marín, F. Ferreres, F. A. Tomás-Barberán and M. I. Gil, *J. Agric. Food Chem.*, 2004, **52**, 3861.
23. N. Baenas, M. Belović, N. Ilic, D. A. Moreno and C. García-Viguera, *Food Chem.*, 2019, **274**, 872–885.
24. T. J. Kim, H. Hyeon, N. Il Park, T. G. Yi, S. H. Lim, S. Y. Park, S. H. Ha and J. K. Kim, *Food Chem.*, 2020, **331**, 127286.
25. S. W. Meckelmann, D. W. Riegel, M. J. Van Zonneveld, L. Ríos, K. Peña, R. Ugas, L. Quinonez, E. Mueller-Seitz and M. Petz, *J. Agric. Food Chem.*, 2013, **61**, 2530–2537.
26. L. S. Ching and S. Mohamed, *J. Agric. Food Chem.*, 2001, **49**, 3101–3105.
27. S. W. Meckelmann, C. Jansen, D. W. Riegel, M. van Zonneveld, L. Ríos, K. Peña, E. Mueller-Seitz and M. Petz, *Eur. Food Res. Technol.*, 2015, **241**, 817–825.

28. Z. Nagy, H. Daood, A. Koncsek, H. Molnár and L. Helyes, *J. Liq. Chromatogr. Relat. Technol.*, 2017, **40**, 199–209.
29. Y. Wahyuni, A. R. Ballester, E. Sudarmonowati, R. J. Bino and A. G. Bovy, *Phytochemistry*, 2011, **72**, 1358–1370.
30. S. Bijttebier, K. Zhani, E. D'Hondt, B. Noten, N. Hermans, S. Apers and S. Voorspoels, *J. Agric. Food Chem.*, 2014, **62**, 4812–4831.
31. R. Fatin Najwa and A. Azrina, *Int. Food Res. J.*, 2017, **24**, 726–733.
32. C. Emmanuel-Ikpeme, *J. Food Nutr. Sci.*, 2014, **2**, 74.
33. K. M. Phillips, D. M. Ruggio, M. Ashraf-Khorassani and D. B. Haytowitz, *J. Agric. Food Chem.*, 2006, **54**, 9998–10002.
34. F. Karatas, A. Baysar and M. Alpaslan, *J. Food Process. Preserv.*, 2017, **41**, 1–5.
35. N. Zaki, A. Hasib, A. Hakmaoui, F. Dehbi and A. Ouatmane, *J. Nat. Sci. Res.*, 2013, **3**, 111–119.
36. L. Gu, H. Pang, K. Lu, H. Liu, X. Wang and G. Qin, *J. Sci. Food Agric.*, 2017, **97**, 1894–1903.
37. G. T. de Souza Sora, A. H. P. Souza, A. A. F. Zielinski, C. W. I. Haminiuk, M. Matsushita and R. M. Peralta, *Cienc. Agrotecnol.*, 2015, **39**, 372–380.
38. K. M. Borges, L. B. O. Vilarinho, A. A. Melo Filho, B. S. Morais and R. N. S. Rodrigues, *Rev. Agron. Ambiente*, 2015, **9**, 292.
39. R. L. Jarret, I. J. Levy, T. L. Potter and S. C. Cermak, *J. Food Compos. Anal.*, 2013, **30**, 102–108.
40. R. Ananthan, K. Subhash and T. Longvah, *Food Chem.*, 2016, **238**, 51–57.
41. C. Taiti, C. Costa, C. A. Migliori, D. Comparini, S. Figorilli and S. Mancuso, *Food Bioprocess Technol.*, 2019, **12**, 1366–1380.
42. A. Korkmaz, A. F. Atasoy and A. A. Hayaloglu, *Food Chem.*, 2020, **311**, 125910.
43. A. Korkmaz, A. A. Hayaloglu and A. F. Atasoy, *LWT–Food Sci. Technol.*, 2017, **84**, 842–850.
44. G. Guclu, D. Keser, H. Kelebek, M. Keskin, Y. Emre Sekerli, Y. Soysal and S. Selli, *Food Chem.*, 2021, **338**, 128129.
45. M. Cirlini, G. Luzzini, E. Morini, S. Folloni, R. Ranieri, C. Dall'Asta and G. Galaverna, *Eur. Food Res. Technol.*, 2019, **245**, 2669–2678.
46. Y. Murakami, H. Iwabuchi, Y. Ohba and H. Fukami, *J. Oleo Sci.*, 2019, **68**, 1251–1260.
47. A. Muscolo, T. Papalia, C. Mallamaci, S. Carabetta, R. Di Sanzo and M. Russo, *Foods*, 2020, **9**, 1323.
48. M. E. Bergman, B. Davis and M. A. Phillips, *Molecules*, 2019, **24**, 1–23.
49. J. Jiang, H. Cen, C. Zhang, X. Lyu, H. Weng, H. Xu and Y. He, *Postharvest Biol. Technol.*, 2018, **146**, 147–154.
50. V. C. Lemos, J. J. Reimer and A. Wormit, *Agriculture*, 2019, **9**, 1–29.
51. A. Rodriguez-Mateos, D. Vauzour, C. G. Krueger, D. Shanmuganayagam, J. Reed, L. Calani, P. Mena, D. Del Rio and A. Crozier, *Arch. Toxicol.*, 2014, **88**, 1803–1853.
52. M. A. S. Coutinho, M. F. Muzitano and S. S. Costa, *Rev. Virtual Quim.*, 2014, **1**, 241–256.

53. J. Oney-Montalvo, A. Uc-Varguez, E. Ramírez-Rivera, M. Ramírez-Sucre and I. Rodríguez-Buenfil, *Agronomy*, 2020, **10**, 1234.
54. A. M. Ribes-Moya, A. M. Adalid, M. D. Raigón, P. Hellín, A. Fita and A. Rodríguez-Burruezo, *J. Sci. Food Agric.*, 2020, **100**, 2208–2223.
55. L. R. Howard, S. T. Talcott, C. H. Brenes and B. Villalon, *J. Agric. Food Chem.*, 2000, **48**, 1713–1720.
56. R. K. Shaha, S. Rahman and A. Asrul, *Ann. Biol. Res.*, 2013, **4**, 27–34.
57. F. Menichini, R. Tundis, M. Bonesi, M. R. Loizzo, F. Conforti, G. Statti, B. De Cindio, P. J. Houghton and F. Menichini, *Food Chem.*, 2009, **114**, 553–560.
58. S. R. Bhandari, B.-D. Jung, H. Baek and Y.-S. Lee, *HortScience*, 2013, **48**, 1275–1282.
59. M. Ghasemnezhad, M. Sherafati and G. A. Payvast, *J. Funct. Foods*, 2011, **3**, 44–49.
60. B. H. Gangadhar, R. K. Mishra, G. Pandian and S. W. Park, *HortScience*, 2012, **47**, 1729–1735.
61. R. L. Jarret, E. Baldwin, B. Perkins, R. Bushway and K. Guthrie, *HortScience*, 2007, **42**, 16–19.
62. R. L. Jarret, T. Berke, E. A. Baldwin and G. F. Antonious, *Chem. Biodiversity*, 2009, **6**, 138–145.
63. J. S. Kim, J. Ahn, S. J. Lee, B. K. Moon, T. Y. Ha and S. Kim, *J. Food Sci.*, 2011, **76**, C193–C198.
64. G. F. Antonious, T. S. Kochhar, R. L. Jarret and J. C. Snyder, *J. Environ. Sci. Health, Part B*, 2006, **41**, 1237–1243.
65. K. D. O. Nascimento, J. Vicente, T. Saldanha, J. Lucena, J. L. Barbosa Júnior and M. I. M. J. Barbosa, *Rev. Verde Agroecol. e Desenvolv. Sustentável*, 2012, 7, 283–288.
66. B. C. Santos, L. E. da Silva, D. A. dos Anjos, B. de, A. B. Mendes, C. C. E. de Souza and M. E. Zanuto, *Braz. J. Dev.*, 2020, **6**, 56214–56231.

CHAPTER 5

Sensory Attributes of Capsicum

5.1 Introduction

When a Carolina Reaper chilli is sliced, everybody in the room will know it. The scent of pepper impregnates the air almost instantly, with its pungency almost incapacitating. The taste can be felt in the air. Even a single, small slice of this deep red pepper can make you feel like you just drank half a litre of a boiling drink. The eyes and nose of everyone in the room will feel like they have been pierced by a sharp and tiny blade.

This description refers to the "sensory features": the characteristics detectable by the human senses. These characteristics include flavour, odour, colour and texture. Almost every mammal uses their senses to choose food. Instinctively, we eat foods that have attractive aromas and colours.[1,2] We also use our senses to identify when food has become rotten. In fact, all these sensory features can be used as a measure to establish the quality of edible products.

Experiencing foods involves all five senses. Have you ever tried to eat with your eyes closed or without touching the food? It is challenging and requires practice to become used to the sensations. We start to perceive a meal from the moment our eyes see it. It must have a good appearance. This is why colour is considered to a quality feature in the analysis of food additives. Colour is produced by the chromogenic compounds present in any object. Any modification in the colour of a food therefore reflects its chemical profile.[3–5]

Sensory features are basically our brain's interpretation of the electrical signals it receives *via* our nerves. We must remember that most of the sensors in our nervous system involve interactions with a specific chemical compound. For instance, the pain receptors that interact with capsaicin to

Food Chemistry, Function and Analysis No. 37
Chemistry and Nutritional Effects of Capsicum
By Valdir Florencio da Veiga, Jr, Larissa Silveira Moreira Wiedemann,
Claudio Pereira de Araujo, Jr and Ananda da Silva Antonio

Published by the Royal Society of Chemistry, www.rsc.org

produce the pungency sensation are cation channels in cellular membranes. The same concept is applicable to the olfactory and gustatory systems.[2,3,6]

Characterising a food by our senses seems easy – after all, we do it daily – but it is more intricate that we are able to describe aloud. Have you ever wondered how you would describe a specific colour within a palette of hundreds of different dark red tones? Have you ever tried to identify and describe a mixture of scents? We do not have a technology like photography to show others a scent or taste, so how can we be sure they understand our description of these features? Sensory features are an extremely complex subject and we do not have methods to describe and record all of them. This chapter comprehensively reviews the sensory characterisation of *Capsicum*.

5.2 Colour

The colours of *Capsicum* are influenced by the accumulation of carotenoids, anthocyanins and chlorophyll. The accumulation of these chemical compounds is defined in the plant's DNA. Their production depends on how the plant grows and experiences abiotic and biotic stressors. The colour of each of the plant's organs during their life cycle can be traced back to their genes. For instance, the colour of unripe fruits is related to genes such as *sw1* (sulphur white), *sw2* (yellowish green) and *sw3* (cedar green). Depending on the species, the fruit will have a certain coloration. Typical colours of *Capsicum* include green, yellow, orange, red and brown (Figure 5.1).[7,8]

In this genus, ripening also causes significant changes in fruit colour (Figure 5.2). The red and yellow fractions of carotenoids (see Chapter 4) have an important role in the definition of colour. As the biosynthesis and accumulation of metabolites in plants is controlled by gene expression, colours can also be associated with specific gene loci (locations) in its DNA. For instance, orange fruits are mainly associated with the *C2* locus and the genes phytoene synthase and capsanthin–capsorubin synthase. Red in fruits is associated with capsanthin–capsorubin synthase in the *Y* locus and genes in the *C1* locus influence the concentration of each capsaicinoid. The gene capsanthin–capsorubin synthase is important in the study of *Capsicum* colours because it is responsible for the conversion of the yellow fraction carotenoids into their red fraction forms. Modification of this gene can determine whether the species will have red, yellow or orange fruits.[9] As a result, *Capsicum* fruits are diverse in colour.

The colours of *Capsicum* bioproducts can be more precisely identified by colorimetry. Colorimeters are simplified spectrophotometers that measure the interaction of the sample with wavelengths corresponding to the colours red, green and blue (RGB), mimicking the human eye. Depending on the transmittance in each spectrum, the colorimeter will describe the sample colour as a mathematical equation in the RGB colour code. The RGB colour code is composed of three numbers representing the brightness (L transitions through black to white), the intensity of red to green (usually represented by a) and the intensity of yellow to blue (usually represented by b) (Figure 5.3).

1
2
3
4
5
6
7
8
9
10
11
12
13
14
15
16
17
18
19
20
21
22
23
24
25
26
27
28
29
30
31
32
33
34
35
36
37
38
39
40
41
42
43
44
45
46
47
48
49
50
51
52
53
54
55
56
57
58
59
60
61
62
63
64
65
66

Figure 5.2 Different colours of *Capsicum* fruits during ripening. Image credit: Carlos Magno da Silva Antonio.

With this type of code, it is possible to mathematically detect small nuances in the colour of a product, an important parameter in the quality control of food.[4] The colour scale most used in food analysis is the CIE (Commission Internationale de l'Eclairage) system, which is a colour-matching system based on the RGB code.[4,10–12]

The colour measurements given by the RGB and CIE systems are also called the "apparent colour" – that is, the colour that can be observed. Other measurements can also be taken to classify *Capsicum* bioproducts, known as the "extractable colour". This coloration is observed after extraction of the sample with acetone, followed by spectrophotometric measurements of the absorbance at 460 nm. The extractable colour analysis method is

Figure 5.1 Examples of *Capsicum* species with different colours. 1: Bhut Jolokia, 2: Trinidad 7 Pot, 3: Trinidad Butch T Scorpion, 4: Trinidad Douglah, 5: Trinidad Moruga Scorpion, 6: Yellow Bhut Jolokia (Ghost Pepper), 7: Aji Crystal, 8: Jalamundo, 9: Manzano, 10: Corno Di Toro, 11: Banana Sweet, 12: Lilac Bell, 13: Succette de Provence, 14: Thai Red, 15: Ancho, 16: Joe Parker, 17: Big Jim, 18: Sandia, 19: Feher Ozon Paprika, 20: Aconcagua, 21: Santa Fe Grande, 22: 5 Color Marble, 23: Twilight, 24: Naga Dorset, 25: Pasilla, 26: Gold Nugget, 27: Sangria, 28: Peach Habanero, 29: Korean, 30: Chilly Chili, 31: Scotch Bonnet, 32: Espanola, 33: Haiti Cluster, 34: Guajillo, 35: Atomic Starfish, 36: Nepal, 37: Spanish Cayenne, 38: Chinese Giant Sweet, 39: Amish Chicken, 40: Bulgarian, 41: Rocoto Yellow, 42: Shishito, 43: Aji Limon, 44: Sunrise, 45: Mayan Cobanero, 46: Szegedi Giant, 47: Marseilles Sweet Yellow, 48: White Habanero, 49: Tepin, 50: Apple Pimento, 51: Dulcetta Orange, 52: Japones, 53: Joe's Long Cayenne, 54: Tequila Sunrise, 55: Uba Tuba, 56: White Bhut Jolokia, 57: Lady Bug Cherry Bomb, 58: Brown Bhut Jolokia, 59: Orange Trinidad Moruga, 60: Chili De Arbol, 61: Big Bertha, 62: Mustard Habanero, 63: Bahama Fish, 64: Barancio Paprika, 65: India Byadagi Mirchi, 66: Gundo Mirchi. Reproduced from ref. 8, https://doi.org/10.1371/journal.pone.0161464, under the terms of the CC BY 4.0 license, https://creativecommons.org/licenses/by/4.0/.

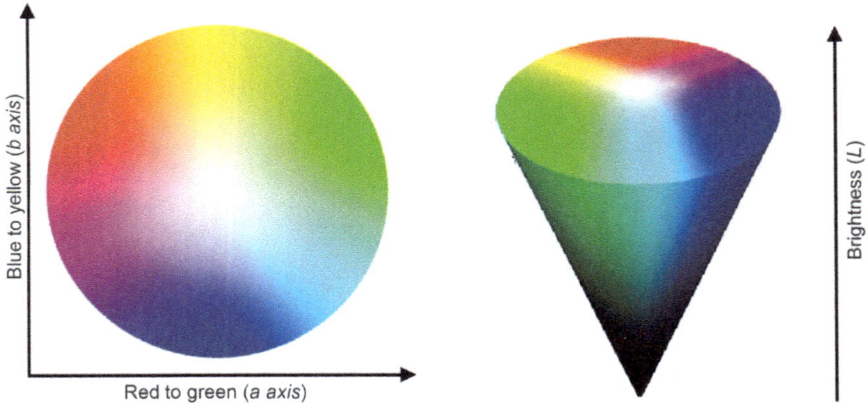

Figure 5.3 RGB system for the representation of colour (the scale number will depend on the mathematical model). Image credit: Carlos Magno da Silva Antonio.

standardised by the American Spice Trade Association (ASTA)[13] and expressed in ASTA units. The quality control for commercialisation uses the ASTA method as an official colour parameter. This method is known as ASTA-20 and is applied for oleoresin and paprika samples. The ASTA-20 method can be summarised in five steps.[5,14–17]

1. Prepare a standard solution of 0.3005 $g\,L^{-1}$ potassium dichromate with 34.96 $g\,L^{-1}$ of ammonium cobalt^{II} sulfate in 1.8 M sulfuric acid. The absorbance of this solution (A_s) at 460 nm is nearly 0.600. This will be used to calculate a correction factor (l_f) for the method.
2. Weigh the sample (50–80 mg) and mix in 100 mL of acetone. Stir the mixture for at least 16 h.
3. Filter and dilute an aliquot of the resulting extract by a factor of 10.
4. Measure the absorbance of the sample (A) at 460 nm.
5. Apply the following equations:

$$l_f = \frac{0.600}{A_s}$$

$$\text{ASTA unit} = \frac{A \times 164 \times l_f}{\text{sample weight (g)}}$$

Depending on the application of the product, the ASTA colour value is acceptable for commercial use within a certain range. In general, higher ASTA units indicates a redder colour and these are usually accepted for food colouring purposes (Figure 5.4). High-quality paprika samples for food colouring usually have an ASTA value >100, indicating that they have a higher content of chromogenic compounds, such as the red fraction carotenoids.[14,15]

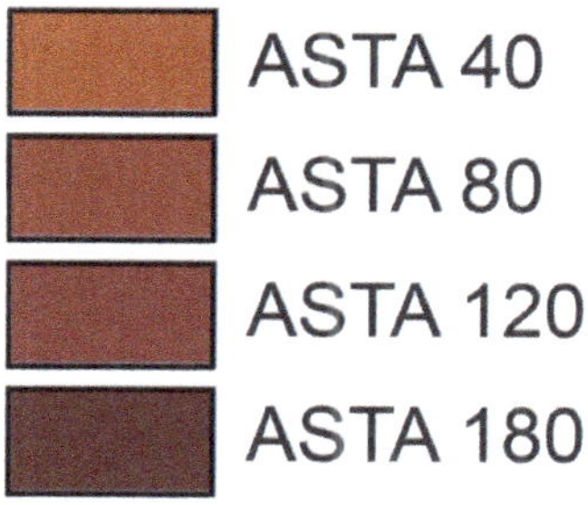

Figure 5.4 Representation of observable colours in paprika samples with ASTA colour units of 40, 80, 120 and 180. Image credit: Carlos Magno da Silva Antonio.

The intensity and redness of a *Capsicum* fruit or any of its bioproducts will be closely influenced by both biotic and abiotic factors, such as solar radiation, soil fertility, the storage conditions, the extraction method, ripening and the metabolomic profile of the species. The intensity of colour is not related to the pungency, but to the content of chromogenic metabolites.[17]

Colour modifications are also used as a tool to classify *in natura* peppers by their maturity stage. This is the tool most used in cultivation fields to determine if a pepper is ready to be harvested. In addition, consumers choose peppers mainly by their appearance. People think that they can perceive whether a fruit is mature by observing its appearance; however, the different colours of *Capsicum* make this difficult. For instance, it is harder to visualise maturity in green cultivars because the colour modifications are too subtle for the human eye.

A recent alternative method used to define maturity without destructive chemical analyses is to analyse photographs. The computational analysis of photographs based on the RGB system and the size of the fruit has been shown to be efficient for the determination of maturity[18,19] and the digital analysis of RGB colour intensity and the digital size of fruits (on a pixel scale) can give a maturity classification system with at least 89% accuracy. This classification system requires calibration for each cultivar in each harvesting season. However, the calibration process requires <50 fruits and this method can be used for the analysis of agricultural data during harvesting.

5.3 Aroma

Aroma and flavour are major factors in making a meal attractive. The aroma of peppers is mainly related to their volatile organic metabolites. However, if you use different peppers for cooking, you have probably noticed that each one has a distinct aroma. Some peppers are pungent, whereas others have sweet or even citric notes (Table 5.1). Some cultivars, although non-pungent, have a strong characteristic aroma, such as orange biquinho, a variety of *Capsicum chinense*.

Table 5.1 Aromas of different *Capsicum* species.

Capsicum species	Aroma
C. annuum jalapeno	Green, earthy, vegetable, pungent
C. annuum serrano	Green, pea-like, fruity, pungent
C. chinense orange biquinho	Green, spicy
C. chinense habanero	Fruity, sweet, spicy
C. frutescens tabasco	Sweet, fruity, pungent, spicy

Variations in the aroma of *Capsicum* are caused by qualitative and quantitative features of their volatile organic compound (VOC) profile. The VOC profile of *Capsicum* mainly consists of esters, terpenes, hydrocarbons, alcohols, aldehydes and ketones. The combinations and relative proportions of the compounds result in a distinct aroma (Figure 5.5). Let us take the orange biquinho and Scotch bonnet chillies as examples, both of which are *C. chinense* varieties. Orange biquinho is known for its low pungency and sweet green smell, whereas Scotch bonnet is one of the most pungent peppers in the world and has a strong and sharp ethereal aroma, also described as a pungent aroma. A common characteristic of the *C. chinense* VOC profile is the occurrence of esters, representing >30% of the total VOC content, with high concentrations of hexyl isopentanoate. The total content of esters in orange biquinho chillies is <23% and hexyl isopentanoate is almost absent. One of the major compounds in orange biquinho chillies is α-cadinol, which is absent in highly pungent *C. chinense* varieties. This small modification in the VOC profile is primarily responsible for the huge difference in the aroma of these two peppers (Figure 5.5).[20–23]

The pepper aroma will be altered from green to sweet, fruity or spicy depending on the qualitative and quantitative profiles of the VOCs. Each chemical class is associated with a distinct scent and influences the final aroma (Table 5.2).[1,17,24] It is possible to explain the distinct aroma between *Capsicum annuum*, *C. chinense* and *Capsicum frutescens* based on 23 volatile compounds.[25] Despite the potential of such characterisation for quality control and systematic issues, and even for breeding new cultivars, there is still little research focused on VOCs compared with capsaicinoids.[1]

5.4 Flavour

When we think about the flavour of peppers, our first thought goes to the pungency. But does pungency really mean flavour? The answer to this question is a little tricky. The perception of flavour is a mixture of sensory signals obtained by several different nerves and receptors in the human body, which include aroma, taste, temperature and texture. If we consider flavour only as the sensations generated by the gustatory system, pungency alone is not flavour because it is caused by the perception of capsaicinoids in the pain and heat receptors of the nervous system. The gustatory system alone only recognises five basic tastes: sweet, salty, sour, bitter and savoury. All other tastes and flavours are a combination of different perceptions of

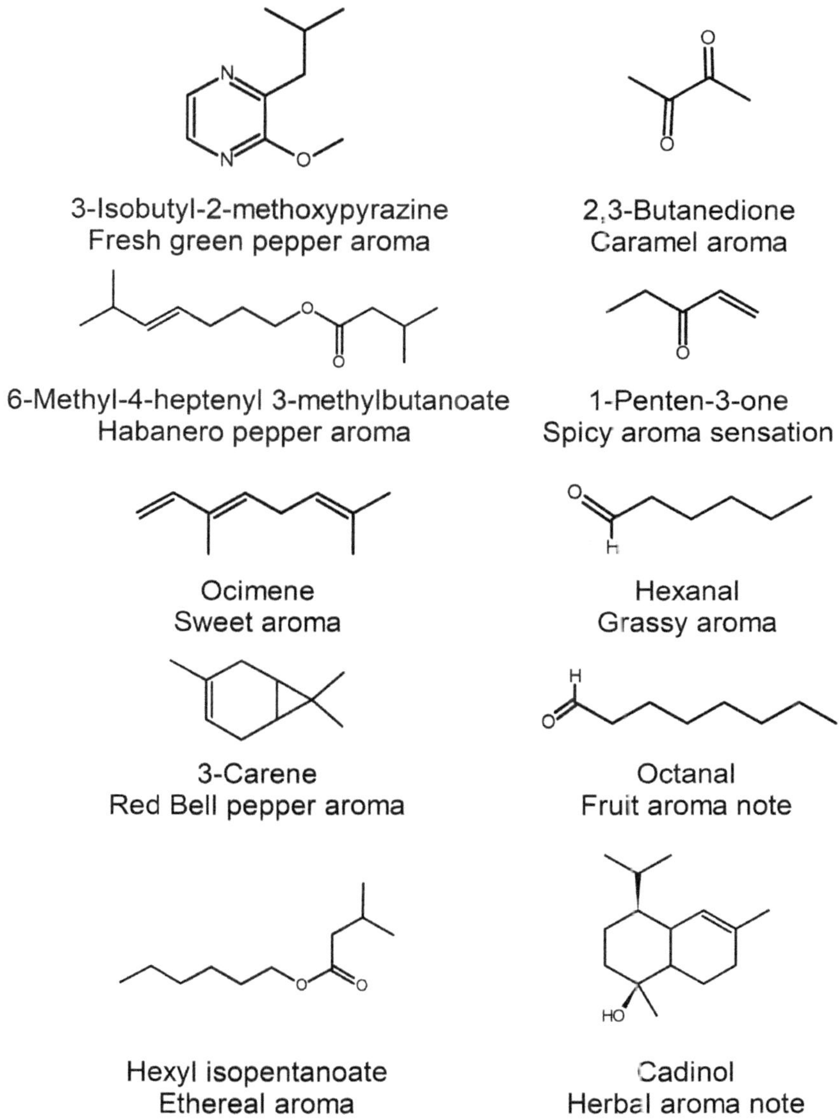

Figure 5.5 Chemical structures of some of the volatile organic compounds present in *Capsicum* fruits that are responsible for aroma. Image credit: Carlos Magno da Silva Antonio.

our senses.[2,3] The flavour of *Capsicum* fruits is a combination of the capsaicinoid content and the VOCs.[26] As we start thinking of flavours as combinations of taste, aroma, temperature and pain, it is easy to note that the flavour of *Capsicum* is also related to the major compounds that cause their distinct aromas. Understanding the metabolites responsible for flavours and

Table 5.2 Aromas perceived in the volatile organic compounds of *Capsicum*.

Volatile organic compound	Aroma
Ethyl 3-methylpentanoate	Fruity, exotic
α-Pinene	Pinewood, herbal
Linalool	Citrus, fruity, floral
Hexanal	Green
D-Limonene	Citrus
Trimethyl pyrazine	Nutty
3-Methylbutanol	Fermented
1-Octen-3-ol	Earthy
2-Methyl propanal	Spicy
Hexyl iso-butanoate	Green
Acetic acid	Acidic
Methyl pentanoate	Pungent
Methyl hexanoate	Ethereal
Acetone	Slightly pungent
Methyl salicylate	Green, sweet
4-Methyl 1-hexanol	Sweaty
Ectocarpene	Sweet
4-Methylpentyl 2-methylbutanoate	Sweet green

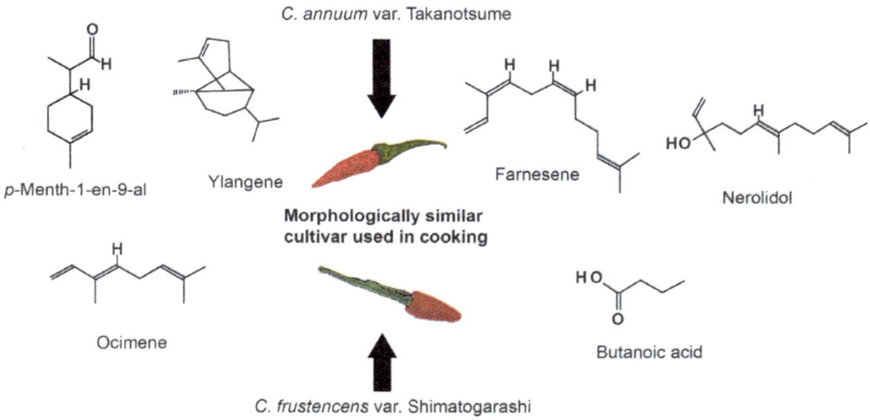

Figure 5.6 Key metabolites in the production of distinct flavours between morphologically similar cultivars. Image credit: Carlos Magno da Silva Antonio.

the mechanisms that generate them can enhance food technology, especially with regard to the breeding of species and storage methods.

However, this type of analysis requires untargeted metabolomic approaches and multivariate chemometric analysis, neither of which are commonly used in this research area. Recent papers have described the major key metabolites contributing to the flavours of popular varieties of *Capsicum* species.[26–28] For instance, the chilli pepper varieties takanotsume and shimatogarashi have a similar fruit appearance, but different nutritional properties (Figure 5.6), making it easier to mistake the identity of both species visually. One of the differences between them is the perceived flavour. The variety takanotsume has more intense fruity and sweet notes, which are associated with higher concentrations of *p*-menth-1-en-9-al, β-ocimene, ylangene, β-farnesene, nerolidol and butanoic acid.[27] Such compounds are useful not only to distinguish among different species, but also between cultivars of the same species.

Other compounds associated with the sweetness in *Capsicum* fruits include (*E*)-2-hexen-1-ol, neopentane, (*E*)-geranylacetone, glucose and fructose. β-Ocimene occurs in nature as two isomers, but only the *E* isomer (Figure 5.6) has a sweet taste. An absence or decrease in sweetness can mainly be correlated with (*Z*)-2-penten-1-ol, the rubber-like smell of which can mask other tastes.[29]

In general, the key flavour metabolites of *C. annuum* are alkenes and terpenes, whereas in *C. chinense* and *Capsicum baccatum* they are esters.[26] A similar pattern as observed for aroma. It is possible to define key metabolites in the flavour of similarly pungent cultivars – for example,[26] the different flavours of pungent accessions of *C. chinense* are correlated with the presence of cyclic hydrocarbons, phenol derivates and norcarotenoids within their esters.

5.5 Concluding Remarks

Capsicum peppers are common in our daily life and are used to prepare meals and as food additives. Their popularity is due to their flavour and aroma. However, such attributes are usually neglected in their scientific characterisation, in part due to the lack of methods to measure such complex traits. As sensory features are caused by hundreds of volatile and non-volatile compounds, it is difficult to define the key factors of an aroma or a flavour. The chemical characterisation of aroma and flavour can aid in the technological development of food-grade bioproducts. Understanding which chemical profiles and biosynthetic routes are correlated with sensorial attributes can aid the development of varieties with higher commercial value. There is plenty of evidence that compounds such as esters, norcarotenoids and alkenes can explain the distinct sensory features of peppers; however, a lot of work is still required to characterise the large quantity of different species, varieties, cultivars and accessions.

References

1. S. Koeda, K. Sato, K. Tomi, Y. Tanaka, R. Takisawa, M. Hosokawa, M. Doi, T. Nakazaki and A. Kitajima, *J. Jpn. Soc. Hortic. Sci.*, 2014, **83**, 244–251.
2. K. Lindblom, *Flavor Chemistry Research at the USDA Western Regional Research Center*, American Chemical Society National Historic Chemical Landmarks, Washington D. C., 2013.

3. I. Guzman, P. W. Bosland and M. A. O. Connell, in *The Biological Activity of Phytochemicals, Recent Advances in Phytochemistry*, ed. D. R. Gang, Springer-Verlag New York, New York, 41st edn, 2011, pp. 109–126.
4. P. B. Pathare, U. L. Opara and F. A. J. Al-Said, *Food Bioprocess Technol.*, 2013, **6**, 36–60.
5. M. Krajayklang, A. Klieber and P. R. Dry, *Postharvest Biol. Technol.*, 2000, **20**, 269–278.
6. A. Korkmaz, A. F. Atasoy and A. A. Hayaloglu, *Food Chem.*, 2020, **311**, 125910.
7. A. Srivastava and M. Mangal, in *The Capsicum Genome, Compendium of Plant Genomes*, ed. N. Ramchiary and C. Kole, Springer Nature Switzerland, Cham, 2019, pp. 25–55.
8. M. B. Kantar, J. E. Anderson, S. A. Lucht, K. Mercer, V. Bernau, K. A. Case, N. C. Le, M. K. Frederiksen, H. C. DeKeyser, Z. Z. Wong, J. C. Hastings and D. J. Baumler, *PLoS One*, 2016, **11**, 1–12.
9. R. L. Jarret, G. E. Barboza, F. R. da, C. Batista, T. Berke, Y. Y. Chou, A. Hulse-Kemp, N. Ochoa-Alejo, P. Tripodi, A. Veres, C. C. Garcia, G. Csillery, Y. K. Huang, E. Kiss, Z. Kovacs, M. Kondrak, M. L. Arce-Rodriguez, M. A. Scaldaferro and A. Szoke, *J. Am. Soc. Hortic. Sci.*, 2019, **144**, 3–22.
10. N. Riquelme and S. Matiacevich, *CYTA-J. Food*, 2017, **15**, 344–351.
11. E. Yilmaz, E. Sevgi Arsunar, B. Aydeniz and O. Güneşer, *Eur. J. Lipid Sci. Technol.*, 2015, **117**, 1016–1026.
12. J. Pino, M. González, L. Ceballos, A. R. Centurión-Yah, J. Trujillo-Aguirre, L. Latournerie-Moreno and E. Sauri-Duch, *Food Chem.*, 2007, **104**, 1682–1686.
13. Food and Agriculture Organization of the United Nations, *Oleoresin paprika*, 2006.
14. N. Zaki, A. Hasib, A. Hakmaoui, F. Dehbi and A. Ouatmane, *J. Nat. Sci. Res*, 2013, **3**, 111–119.
15. N. Zaki, A. Hakmaoui, A. Ouatmane and J. P. Fernandez-Trujillo, *Food Sci. Technol.*, 2013, **33**, 577–585.
16. B. H. Gangadhar, R. K. Mishra, G. Pandian and S. W. Park, *HortScience*, 2012, **47**, 1729–1735.
17. M. Cirlini, G. Luzzini, E. Morini, S. Folloni, R. Ranieri, C. Dall'Asta and G. Galaverna, *Eur. Food Res. Technol.*, 2019, **245**, 2669–2678.
18. B. Harel, Y. Parmet and Y. Edan, *Comput. Ind.*, 2020, **121**, 103274.
19. M. J. Villaseñor-Aguilar, M. G. Bravo-Sánchez, J. A. Padilla-Medina, J. L. Vázquez-Vera, R. G. Guevara-González, F. J. García-Rodríguez and A. I. Barranco-Gutiérrez, *Appl. Sci.*, 2020, **10**, 5097.
20. L. F. Cuevas-Glory, O. Sosa-Moguel, J. Pino and E. Sauri-Duch, *Food Anal. Methods*, 2015, **8**, 1005–1013.
21. A. Gahungu, E. Ruganintwali, E. Karangwa, X. Zhang and D. Mukunzi, *Adv. J. Food Sci. Technol.*, 2011, **3**, 211–218.
22. V. C. Castro Alves, N. O. F. Pinto, M. F. A. Penha, B. L. Gomes, F. J. B. Reifschneider and D. S. Garruti, in *Flavour Science*, ed. V. Ferreira and R. Lopez, Academic Press, London, 2014, pp. 567–571.

23. J. A. Pino, E. Sauri-Duch, O. Sosa-Moguel, C. A. Can-Cauich, V. M. Moo-Huchin and L. Cuevas-Glory, *Acta Aliment.*, 2020, **49**, 321–329.
24. Manikharda, M. Takahashi, M. Arakaki, K. Yonamine, F. Hashimoto, K. Takara and K. Wada, *J. Oleo Sci.*, 2018, **67**, 113–123.
25. A. Rodriguez-Burruezo, C. Gonzalez-Mas Mdel and F. Nuez, *J. Food Sci.*, 2010, **75**, S446–S453.
26. E. Morales-Soriano, B. Kebede, R. Ugás, T. Grauwet, A. Van Loey and M. Hendrickx, *Food Res. Int.*, 2018, **109**, 250–259.
27. Manikharda, M. Takahashi, M. Arakaki, K. Yonamine, Y. Asikin, K. Takara and K. Wada, *Food Sci. Technol. Res.*, 2017, **23**, 427–435.
28. Y. Song, B. Du, Z. Ding, Y. Yu and Y. Wang, *LWT*, 2021, **139**, 110525.
29. P. M. Eggink, C. Maliepaard, Y. Tikunov, J. P. W. Haanstra, A. G. Bovy and R. G. F. Visser, *Food Chem.*, 2012, **132**, 301–310.

CHAPTER 6

Nutritional Aspects of Capsicum

6.1 Introduction

Capsicum peppers are one of the oldest and most important food colourants and spices in the world. They are widely present in the cuisines of different cultures, being prepared and cooked as a vegetable to compose different dishes, including salads. Chillies and sweet peppers have been used in the human diet since ancient times, from at least 5000 years BC. However, the popularity of *Capsicum* is not only related to its unique sensorial features, but also to its nutritional and health-promoting properties.

Throughout the history of *Capsicum* exploitation by humans, peppers have been used to treat different diseases, including inflammatory conditions, infections, neurological problems, respiratory disorders such as asthma and as pain relief.[1] "Let food be thy medicine, and let medicine be thy food". This famous quote, usually attributed to Socrates, is suitable to describe *Capsicum*. *Capsicum* plants were used as medicine and are still a food that can improve health just by being part of a dish. *Capsicum* fruits are rich in antioxidants, vitamins, minerals, lipids and carbohydrates (Figure 6.1), all of which are considered important in the health of humans and other mammals.[2] The essential requirements to maintain mammalian health and proper functioning (*e.g.* growth, energy production and reproduction) are related to micro- and macronutrients that are not biosynthesised in sufficient amounts in the body.

Nutrition through a healthy diet is an effective preventive method for non-transmissible chronic diseases, including marasmus, osteoporosis, strokes and obesity (Figure 6.2). Peppers are considered to be a reliable and balanced source of essential macro- and micronutrients for good functioning of the mammalian organism, not only due to their macro- and micronutrient

Food Chemistry, Function and Analysis No. 37
Chemistry and Nutritional Effects of Capsicum
By Valdir Florencio da Veiga, Jr, Larissa Silveira Moreira Wiedemann,
Claudio Pereira de Araujo, Jr and Ananda da Silva Antonio

Published by the Royal Society of Chemistry, www.rsc.org

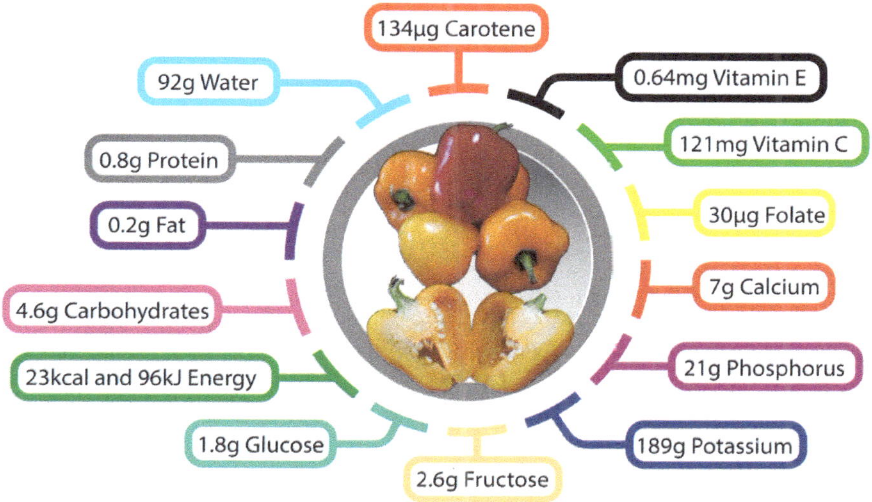

Figure 6.1 Nutritional content of *Capsicum* sweet yellow pepper. Image credit: Carlos Magno da Silva Antonio.

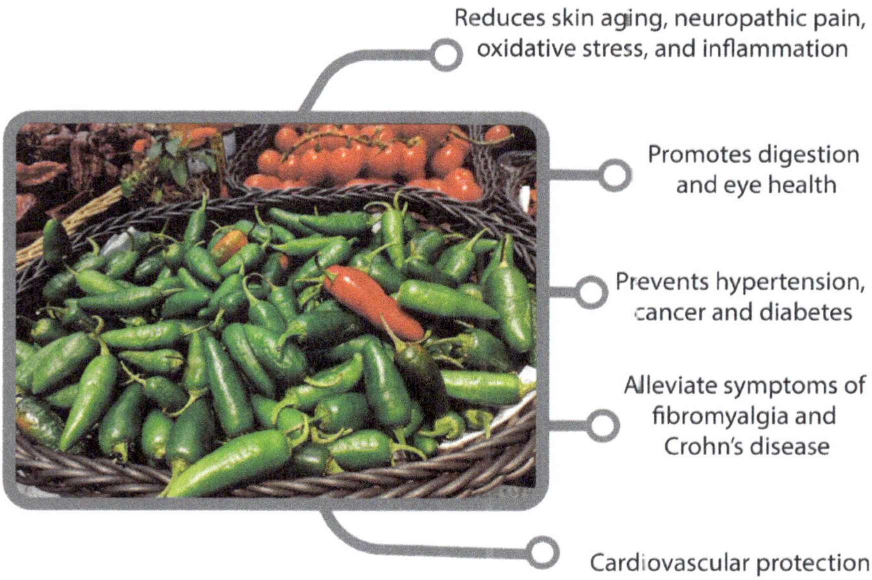

Figure 6.2 Health benefits of the inclusion of *Capsicum* in the diet. Image credit: Carlos Magno da Silva Antonio.

content, but also because of their regular inclusion in cooking. The consumption and production of *Capsicum* species is dependent on the culture of the people – for instance, the per capita consumption of chilli peppers in urban areas of Mexico is 8–9 kg per year.[3]

Studies conducted in Chinese and Italian populations have shown that adherence to a Mediterranean diet including a daily intake of *Capsicum* can reduce the risk of cardiovascular disease by 14–50%.[4–6] Although not all the biological mechanisms involved in the health benefits of *Capsicum* have been elucidated, their nutritional effects are based on their remarkable composition.

6.2 Proximate Composition

One method used to evaluate the nutritional content of food products is the proximate composition. These data are used by nutrition and health professionals to prescribe specific diets. In many countries, it is mandatory to specify proximate composition data on packing. The proximate composition includes five elements: moisture, ash, lipids, protein and carbohydrates. These data summarise the nutraceutical benefits of the product and enable the energy and digestibility of the product to be addressed. The values are usually expressed as the content in 100 g or 100 mL, which is considered to be a single portion (Figure 6.3).[7]

The moisture represents the water content of the product, which, in addition to being an essential part of the mammalian diet, also indicates the shelf life. In general, a high moisture content is indicative of a shorter shelf life because the product is more susceptible to the development of microbial growth. The moisture content of *in natura* fruits is usually higher than that of industrialised or powdered products (Table 6.1). The moisture content of powdered products, such as paprika and other dried condiments, should be <10% to guarantee product quality.[8,9]

The ash content is determined after the complete removal of any moisture and organic material at about 550 °C. The ash content is an indirect measure of the mineral content of the food, with a higher ash content indicating more minerals. The ash content does not provide any information about the mineral profile.[16] The ash content of *Capsicum* (Table 6.1) is significantly higher than other common fruits such as bananas (3.87%), lemons (5.09%), oranges (3.13%) and watermelon (3.48%).[17] These values are strongly indicative of the potential of *Capsicum* to alleviate deficiencies of mineral nutrients.[18]

The importance of protein is related to the development of the mammalian organism: all cells are composed of proteins. They are present in the molecular structure of cellular membranes, enzymes, hair, collagen and serum albumin. They are also precursors of hormones and coenzymes, which are essential to sustain biological function. Without a source of protein, or, more accurately, without a source of amino acids to produce new proteins, the body can collapse. The recommended dietary allowance (RDA) of protein for an adult human to ensure good health is about 0.80 g per kg body weight per day.[19] This daily protein requirement comes from every meal, there is no need to eat several peppers every day. For an adult weighing 60 kg, the consumption of 100 g of rocoto pepper (Table 6.1) (nearly one fruit) contributes about 30% of the daily RDA of protein.

Sweet pepper sauce

Nutritional Facts (serving size 15g)

Amount per serving		%DV
Calories	50kcal or 210 kJ	2%
Carbohydrates	13g	4%
Sodium	221g	9%

%DV - % daily value

Sweet paprika

Nutritional Facts (serving size 5g)

Amount per serving		%DV
Calories	20kcal or 8.6kJ	1%
Carbohydrates	2.7g	1%
Proteins	0.7g	1%
Total fat	0.6% g	1%
Total Fibre	1.7g	1%

%DV - % daily value

Figure 6.3 Example of commercial *Capsicum* bioproducts with incomplete proximate composition tables. Image credit: Carlos Magno da Silva Antonio.

Table 6.1 Proximate composition of *Capsicum* fruits and products (%).

Species or bioproduct	Moisture	Ash	Protein	Fat	Carbohydrate	Ref.
C. annuum guajillo	9.1	7.53	12.89	12.43	58.00	10
C. annuum jalapeño	10.13	7.32	14.36	4.23	63.97	10
C. annuum serrano	11.23	5.81	12.78	2.26	67.93	10
C. annuum de arbol	5.65	8.82	12.75	13.38	59.41	10
C. annuum green bell pepper	92.56	0.3	1.05	0.38	4.98	11
C. chinense habanero	13.01	7.51	13.52	4.63	61.33	10
C. frutescens	10.14	16.67	11.97	0.35	58.81	12
C. pubescens rocoto rojo	92	9.34	15.54	1.76	55.29	13
C. baccatum cambuci (*in natura*)	86.60	0.14	0.95	0.14	9.10	14
C. baccatum cambuci (jam)	42.53	0.81	1.61	0	55.05	14
Paprika	11.2	7.74	14.1	12.9	54	15

Fat is related to the energetic potential of food and also the transport and absorption of fat-soluble vitamins (*e.g.* vitamin A), although there is no recommended daily intake of total fat. Total fat includes all types of fatty acids, saturated and unsaturated, *cis* and *trans*. *Capsicum* species are usually richer in the unsaturated fatty acids (see Chapter 4), aiding the balance of low- and high-density lipoprotein cholesterol (related to the occurrence of coronary heart disease).[19] The fat content is affected by genetic traits of the pepper and each species and cultivar presents a distinct total fat content. The season and environmental conditions also influence the total fat content.[20] A green bell pepper, one of the most consumed *Capsicum* fruits worldwide, has a similar fat content to banana (0.27%), mango (0.29%) and tamarind (0.36%).[17]

Proteins, fats and carbohydrates are the energy sources of food. Carbohydrates are found in larger concentrations than the other two in *Capsicum* (Table 6.1), acting as the main energy resource. The total carbohydrates are related to sugars (Chapter 4) and starches and are essential to good functioning of the brain. The RDA of carbohydrates is 130 g per day, although the estimated average consumption varies from 180 to 230 g per day.

6.3 Micronutrients

Micronutrients are defined as substances that are required in only small amounts by an organism. For instance, the recommended intake of micronutrients in adult humans is less than half that of the macronutrients. Micronutrients include minerals (*e.g.* calcium, potassium, sodium, iron and zinc), water-soluble vitamins (*e.g.* vitamins B1, B2, B6, B12, C and folic acid) and fat-soluble vitamins (*e.g.* vitamins A, D, E and K).[21]

A deficiency of micronutrients leads to several distinct health problems. A fairly common disease caused by micronutrient deficiency is anaemia. This clinical condition affects 33% of the global population and is caused by a lack of iron, folic acid and vitamins B12 and A. Symptoms include a reduced number of red blood cells, fatigue, shortness of breath and dizziness. Iron deficiency can cause alopecia, digestive disorders and headaches.[22] Vitamin deficiencies are commonly associated with fatigue, but the clinical effects of undernutrition can be severe. Vitamin A deficiency is one of the main causes of blindness due to excessive dryness of the ocular globe and can also worsen infections and inflammatory disease. Deficiencies of water-soluble vitamins are related to atherosclerosis, cancer progression, digestive problems, cerebral injuries and liver disorders.[23,24]

6.3.1 Minerals

Mineral nutrients are inorganic compounds that help to regulate electrostatic equilibrium within the mammalian organism. For instance, Ca^{2+} regulates osmosis through cell membranes and forms hard tissues, such as bones and teeth. The chemical composition of specialised metabolites in peppers varies across species and cultivars. The same phenomenon occurs with the mineral content and each mineral responds differently to the abiotic and biotic stressors experienced by the plant.

Kim *et al.* showed that the calcium, magnesium, phosphorus, potassium, sulphate and manganese content of peppers are affected more by abiotic stress (*e.g.* soil fertility) than by biotic factors (*e.g.* different cultivars).[20] Most of the minerals used by plants are obtained from the soil. To give high cultivation yields, crops must be grown in soils with an adequate supply of nutrients, which is determined by the fertilisation method. Several methods of fertilisation can be used, such as organic or mineral fertilisation, fertigation and biofertilisation.[25–28] The best fertilisation schedule to increase the yield and nutrition content of *Capsicum* crops varies with the species/cultivar because each one has different nutritional requirements. Agencies such as the Brazilian Agricultural Research Corporation (EMBRAPA) and the Food and Agriculture Organization of the United Nations recommend specific fertilisation levels based on organomineral products.

The mineral content of peppers varies depending on the species and cultivar (Table 6.2). In general, *Capsicum* are rich in potassium, providing 7–18% of the daily RDA. Cayenne peppers, for instance, can provide 41% of the daily RDA for potassium, 38% of the daily RDA for magnesium, 97% of the daily RDA for iron, 14% of the daily RDA for calcium and 13% of the daily RDA for phosphorus in a single meal. A single pepper cannot fulfil the entire daily RDA of minerals alone, but the addition of peppers to the diet is highly recommended as a good nutritional additive to alleviate mineral deficiencies.[18,29]

6.3.2 Water-soluble Vitamins

Water-soluble vitamins do not accumulate in the human body and must be consumed daily. Any deficiency in these vitamins can cause severe chronic health issues and even mortality. Vitamins C and B are the most well-known of the water-soluble vitamins. Many people have already heard that they should take vitamins C and B to stay healthy and prevent viral infections

Table 6.2 Mineral composition of *Capsicum* fruits (mg 100 g^{-1} of dry fruit).

Species	Ca	Mg	P	K	Fe	Na	Zn	Ref.
C. annuum	114	137.9	292.2	282.1	6.1	26.9	2.1	20
Sweet green pepper	10	10	20	175	34	3	0.3	15
Sweet yellow pepper	11	12	24	212	0.46	2	0.17	15
C. annuum cayenne	148	152	293	201.4	7.8	30	0.3	15
C. annuum jalapeño	12	15	26	248	0.25	3	0.14	15
C. annuum serrano	11	22	40	305	0.86	10	0.26	15
C. frutescens	106	236	31.4	331	67.2	1.17	32.2	12
C. chinense biquinho	16.4	26.6	24.6	351.7	0.5	1.9	0.1	30
C. baccatum dedo de moça	25.8	37.8	40.6	397.4	0.7	2.7	0.2	30
Recommended dietary allowance (RDA) (mg per day)	1000	400	700	1.5[a]	8	4.7	11	31

[a]g per day.

such as flu. Despite the severe health issues that can be caused by a deficiency of these vitamins, their ingestion requires the simple inclusion of vegetables in the diet. For instance, the recommended daily intake of vitamin C for an adult varies from 60 to 200 mg per day.[32,33] A fresh *Capsicum* pepper contains an average of 50–300 mg of vitamin C (depending on the variety and size of the fruit).

A single pepper in the daily diet can nearly match the daily recommended intake of vitamin C. The vitamin C contents of 100 g of fresh fruit for various *Capsicum* species are: *Capsicum annuum*, 152 mg; *Capsicum frutescens*, 174 mg; *Capsicum pubescens*, 221 mg; and *Capsicum chinense*, 107 mg.

A deficiency of the vitamin B complex can disrupt the neural and circulatory systems and cause anaemia. The dietary reference intakes of vitamins B1 (thiamine), B2 (riboflavin), B3 (niacin) and B6 (pyridoxine) for an adult man (aged 30 years) are 1.0, 1.1, 12 and 1.1 mg per day, respectively, and 2.0 μg per day for vitamin B12 (cobalamin).[21] Despite the fact that *Capsicum* can provide a significant amount of vitamin B complex in the edible portions (Table 6.3), there are few data for their content of this vitamin complex.[2,18]

Most of available data are from the US Department of Agriculture National Nutrient Database, which has not been updated since 2010; some of their standard reference values date back 1977. In such a complex genus, with the constant development of new cultivars and adaptive processes, these data are unreliable for describing the current chemical profile of *Capsicum* species.

6.3.3 Fat-soluble Vitamins

The fat-soluble vitamins (A, D, E and K) are responsible for a number of important physiological functions in the human body. They can be obtained from several different types of food, usually with a high content of fatty acids, such as fish oils, sunflower oil, peanuts and vegetables. Vitamin D it also obtained by exposure of the skin to sunlight. The dietary reference intakes of these vitamins for a 30-year-old adult man are 625 μg per day for vitamin A,[21] 12 mg per day for vitamin E,[21] 120 μg per day for vitamin K[31] and 5 μg per day for vitamin D.[31] A deficiency of these vitamins is associated with the several clinical conditions, including: type 2 diabetes, immune

Table 6.3 Concentrations of the vitamin B complex in *Capsicum* bioproducts (mg 100 g^{-1}).[a]

Sample	Thiamine	Riboflavin	Niacin	B6	B12	Ref.
SR *C. annuum* red spices	0.328	0.919	8.7	2.45	0	15
SR *C. annuum* japaleño	0.04	0.07	1.28	0.419	0	15
SR *C. annuum* serrano	0.054	0.081	1.54	0.505	0	15
SR *C. annuum* paprika	0.33	1.23	10.1	2.14	0	15
SR *C. frutescens*	0.09	0.09	0.95	0.278	0	15
C. annuum jalapeño	0.14	0.06	1.12	0.51	0	30
C. baccatum dedo de moça	0.07	0.09	1.24	0.50	0	30

[a]SR = US Department of Agriculture National Nutrient Database for Standard Reference, Legacy.

Table 6.4 Concentrations of fat-soluble vitamin in various *Capsicum* species.[a]

Species	Vitamin A (mg RAE 100 g^{-1})	Vitamin E (α-tocopherol) (mg 100 g^{-1})	Vitamin K (phylloquinone) (μg 100 g^{-1})	Ref.
C. annuum banana	0.102	0.69	9.5	15
C. frutescens	0.354	0.69	14.3	15
C. annuum jalapeño	0.324	3.58	18.5	15
C. annum serrano	0.281	0.69	11.8	15
C. annum bell pepper (green)	0.111	0.37	7.4	15
C. chinense Scotch bonnet	6.045	nd	nd	37
C. annuum var. *grossum*	1.2	54.7	nd	38
C. frutescens var. *baccatum*	1.2	54.3	nd	38

[a]RAE = retinol activity equivalent. nd = not described.

system disorders and blindness (vitamin A); an increase in cell oxidative stress, peripheral neuropathy, ataxia and anaemia (vitamin E); osteoporosis and cardiovascular disease (vitamin K); and susceptibility to infectious disease, bone weaking, hypertension and malfunctioning of homeostasis (vitamin D). Despite several sources of fat-soluble vitamins being available in a healthy diet, the prevalence of deficiencies of these vitamins is alarming, mainly in low- and middle-income countries.[34–36]

Capsicum species are considered to be a reliable functional food and have a considerable content of fat-soluble vitamins compared with the recommended daily intake (Table 6.4). Some research is focused on breeding species with a higher content of these vitamins. Most published studies have focused on provitamin A (a carotenoid). The lack of development in breeding vitamin-enriched varieties can be seen by comparing the data from standard reference *Capsicum* species (the first five entries in Table 6.4) with more recently developed varieties, such as Scotch bonnet and grossum. The use of *Capsicum* species as a source of vitamins (not provitamins) in the diet should be carefully evaluated because both water-soluble and fat-soluble vitamins are thermosensitive and can undergo degradation during post-harvesting procedures and cooking.[37]

6.4 Fibre

Fibre consists of a mixture of non-digestible carbohydrates that can be classified as dietary and functional fibre. Dietary fibre is further defined as a mixture of carbohydrates and lignans in which the plant matrix is nearly intact. Dietary fibre includes non-starch polysaccharides (*e.g.* cellulose, pectin and gums) and plant carbohydrates that are not recovered *via* precipitation with alcohol (*e.g.* inulin). Functional fibres are a mixture of isolated or extracted fibres, including isolated carbohydrates from a plant, animal or synthetic source, such as resistant starch, pectin, chitin, chitosan, inulin, polydextrose and dextrin. The total fibre content is defined as the sum of dietary and functional fibre. It is important to emphasise that the fibre content is sometimes described as crude fibre. Crude fibre is the insoluble part of dietary fibre, which cannot be fermented in the digestive tract.[19] The nutritional benefits of fibre include several physiological effects, such as the sensation of satiety, because fibre is not fully digested. The presence of a layer of fibre in the intestines slows down digestion and the absorption of nutrients and glucose.[19,39,40]

Capsicum peppers are considered to be a good source of dietary fibre and contain at least 1 g of dietary fibre per 100 g of fruit.[18] The dietary reference intake of fibre for an adult varies from 12 to 18 g per day, whereas the reference intake for total fibre varies from 25 to 38 g per day.[19] A statement of "at least 1 g of dietary fibre" for peppers may seem like a low starting value, but do not forget the huge biodiversity with the *Capsicum* genus. The fibre content, like any other chemical compound within the genus, is affected by several factors, such as the genotype and phenotype (different species, varieties and cultivars), the environmental conditions, the cropping conditions, the pre- and post-harvesting procedures, the plant organ and the plant's life cycle.[27,41] For example, Brazilian chilli peppers can have a fibre content varying from 3.6 to 15.9 g total fibre per 100 g of fruit.[2] It is even possible to find varieties with higher total fibre content per 100 g of fruit within *C. annuum*, such as: *C. annuum* var. *abbreviatum*, 20.64 g;[38] *C. annuum* var. *acuminatum*, 28.39 g;[38] *C. annuum* var. BST, 19.8 g;[20] *C. annuum* var. MCB 20.6 g;[20] and *C. annuum* var. GCH, 21.5 g.[20]

6.5 Concluding Remarks

Capsicum can provide benefits for human nutrition, but the extent of these benefits is not fully clear and requires further evaluation. There are few studies characterising the most common nutritional data for *Capsicum* peppers, such as the proximate composition, mineral content and fatty acid profile. Although these attributes may not seem as interesting as the capsaicinoid content, they are essential to fully comprehend the nutritional functionality of peppers. For instance, the water content in the proximate composition can be used to evaluate the shelf life of *Capsicum* bioproducts. Other features, such as the mineral content, pH and ash content can indicate the maturation stage during harvesting and provide insights into the quality of the fruit. The profile of such physical–chemical characteristics will enable the development of bioproducts with specific desired attributes, such as acidity and energetic value. In addition, characterising the nutritional value of *Capsicum* will help nutritionists to establish the right amount of the fruit to include in the diet to enhance and use all the potential of this functional food.

References

1. B. K. Saleh, A. Omer and B. Teweldemedhin, *MOJ Food Process. Technol.*, 2018, **6**, 325.
2. E. R. do Rêgo, F. L. Finger and M. M. do Rêgo, in *Peppers: Nutrition, Consumption and Health*, ed. M. A. Salazar and J. M. Ortega, Nova Science Publishers, New York, 2012, ch. 7, pp. 18–37.

3. A. M. Vera-Guzmán, E. N. Aquino-Bolanos, E. Heredia-García, J. C. Carrillo-Rodríguez, S. Hernández-Delgado and J. L. Chávez-Servia, in *Flavonoids - From biosynthesis to human health*, ed. G. C. Justino, InTech, Rijeka, 2017, ch. 19, p. 404.
4. M. Bonaccio, A. Di Castelnuovo, S. Costanzo, E. Ruggiero, A. De Curtis, M. Persichillo, C. Tabolacci, F. Facchiano, C. Cerletti, M. B. Donati, G. de Gaetano, L. Iacoviello, A. Bonanni, F. Gianfagna, T. Di Prospero, J. Vermylen, I. De Paula Carrasco, S. Costanzo, A. Spagnuolo, D. Assanelli, V. Centritto, M. Olivieri, A. Gialluisi, F. Gianfagna, S. Magnacca, B. Izzi, A. Marotta, F. Noro, F. De Lucia, F. Bracone, S. Dudiez, L. Rago and T. Panzera, *J. Am. Coll. Cardiol.*, 2019, **74**, 3139–3149.
5. J. D. Spence, *J. Am. Coll. Cardiol.*, 2019, **74**, 3150–3152.
6. A. Moseley, K. Laipply and J. Rubinstein, *J. Am. Coll. Cardiol.*, 2020, **75**, 1865.
7. H. Greenfield and D. A. T. Southgate, *Food Composition Data*, Food and Agriculture Organization of The United Nation, Rome, 2nd edn, 2003, p. 295.
8. G. Speranza, R. Lo Scalzo, C. F. Morelli, M. Rabuffetti and G. Bianchi, *J. Food Biochem.*, 2019, **43**, 1–12.
9. R. Tundis, M. R. Loizzo, F. Menichini, M. Bonesi, F. Conforti, D. De Luca and F. Menichini, *Food Res. Int.*, 2012, **45**, 170–176.
10. L. Orellana-Escobedo, L. E. Garcia-Amezquita, G. I. Olivas, J. J. Ornelas-Paz and D. R. Sepulveda, *CYTA-J. Food*, 2013, **11**, 179–184.
11. G. Melgar-Lalanne, A. J. Hernández-Álvarez, M. Jiménez-Fernández and E. Azuara, *Food Bioprocess Technol.*, 2017, **10**, 51–76.
12. C. Emmanuel-Ikpeme, *J. Food Nutr. Sci.*, 2014, **2**, 74.
13. B. A. V. Dominguez, PhD thesis, Universidad Peruana Cayetano Heredia, 2016.
14. K. D. O. Nascimento, J. Vicente, T. Saldanha, J. Lucena, J. L. Barbosa Júnior and M. I. M. J. Barbosa, *Rev. Verde Agroecol. e Desenvolv. Sustentável*, 2012, 7, 283–288.
15. Composition of Standard Reference Capsicum products of the National Nutrient Database for Standard Reference, https://fdc.nal.usda.gov/fdc-app.html#/?query=capsicum (accessed August 2021).
16. Y. Pomeranz and C. E. Melona, in *Food Analysis*, Springer, Boston, 3rd edn, 1994, ch. 35, pp. 602–624.
17. L. E. Garcia-Amezquita, V. Tejada-Ortigoza, E. Heredia-Olea, S. O. Serna-Saldívar and J. Welti-Chanes, *J. Food Compos. Anal.*, 2018, **67**, 77–85.
18. T. L. Olatunji and A. J. Afolayan, *Food Sci. Nutr.*, 2018, **6**, 2239–2251.
19. Institute of Medicine, *Dietary Reference Intakes for Energy, Carbohydrate, Fiber, Fat, Fatty Acids, Cholesterol, Protein, and Amino Acids*, The National Academies Press, Washington D. C., 2005.
20. E. H. Kim, S. Y. Lee, D. Y. Baek, S. Y. Park, S. G. Lee, T. H. Ryu, S. K. Lee, H. J. Kang, O. H. Kwon, M. Kil and S. W. Oh, *Appl. Biol. Chem.*, 2019, **62**, 48.
21. J. Dwyer, in *Encyclopedia of Food Sciences and Nutrition*, ed. B. Caballero, Academic Press, Cambridge, 2nd edn, 2003, pp. 1863–1868.

22. S. R. Pasricha, J. Tye-Din, M. U. Muckenthaler and D. W. Swinkels, *Lancet*, 2021, **397**, 233–248.
23. J. Lykstad and S. Sharma, *StatPearls*, 2019, pp. 5–7, PMID: 30860745.
24. Committee on Diet and Health, *Diet and health: implications for reducing chronic disease risk*, National Academy Press, Washington, D.C., 1989, vol. 27.
25. S. Altintas, *Afr. J. Biotechnol.*, 2012, **11**, 6470–6475.
26. F. S. Albuquerque, Ê. F. Silva, E. Bezerra Neto, A. E. Souza and A. N. Santos, *Hortic. Bras.*, 2012, **30**, 681–687.
27. L. A. Macedo, L. L. de Oliveira, A. F. Leão, A. A. A. Campos, B. de, O. Ferreira, A. C. França and L. T. Sardinha, *Braz. J. Dev.*, 2020, **6**, 70374–70386.
28. E. S. Omar, A. A. A. Gabal, A. A. Alkharpotly and F. I. Radwan, *J. Adv. Agric. Res.*, 2019, **23**, 402.
29. M. A. Alam, M. Saleh, G. M. Mohsin, T. A. Nadirah, F. Aslani, M. M. Rahman, S. K. Roy, A. S. Juraimi and M. Z. Alam, *J. Food Process. Preserv.*, 2020, **44**, 1–11.
30. C. M. F. Pinto, C. L. de, O. Pinto and S. M. L. S. M. L. Donzeles, *Rev. Bras. Agropecuária Sustentável*, 2013, **3**, 108–120.
31. R. M. Padovani, J. Amaya-Farfán, F. A. B. Colugnati and S. M. Á. Domene, *Rev. Nutr.*, 2006, **19**, 741–760.
32. T. S. Agostini-Costa, I. S. Gomes, L. A. M. P. de Melo, F. J. B. Reifschneider and C. S. da Costa Ribeiro, *J. Food Compos. Anal.*, 2017, **57**, 73–79.
33. R. K. Dubey, V. Singh, G. Upadhyay, A. K. Pandey and D. Prakash, *Food Chem.*, 2015, **188**, 119–125.
34. A. A. Albahrani and R. F. Greaves, *Clin. Biochem. Rev.*, 2016, **37**, 27–47.
35. D. K. Dror and L. H. Allen, *Food Nutr. Bull.*, 2011, **32**, 124–143.
36. K. Amrein, M. Scherkl, M. Hoffmann, S. Neuwersch-Sommeregger, M. Köstenberger, A. Tmava Berisha, G. Martucci, S. Pilz and O. Malle, *Eur. J. Clin. Nutr.*, 2020, **74**, 1498–1513.
37. M. B. Kantar, J. E. Anderson, S. A. Lucht, K. Mercer, V. Bernau, K. A. Case, N. C. Le, M. K. Frederiksen, H. C. DeKeyser, Z. Z. Wong, J. C. Hastings and D. J. Baumler, *PLoS One*, 2016, **11**, 1–12.
38. T. L. Olatunji and A. J. Afolayan, *J. Veg. Sci.*, 2020, **26**, 190–207.
39. G. A. Nevara, S. Kharidah, S. Muhammad, N. Zawawi, N. A. Mustapha and R. Karim, *Foods*, 2021, **10**, 754.
40. P. Evenepoel and B. K. Meijers, *Kidney Int.*, 2012, **81**, 227–229.
41. A. A. C. Chaves and S. C. Furtado, *Nutr. Bras.*, 2017, **16**, 194–203.

CHAPTER 7

Capsicum *Pharmacokinetics and Pharmacodynamics*

7.1 Introduction

During the development and evolution of medicine and pharmaceutical drugs, natural products have always had a major role as a framework for the discovery of compounds that can provide the desired therapeutic effects. Even with the current availability of synthetic drugs, natural products are still used in some instances because they are associated with lower side-effects, lower toxicity and are more pleasant to ingest. After all, most people prefer to eat a colourful functional meal than several different pills. It is therefore essential to understand the true pharmacological effects, pharmacodynamics and pharmacokinetics of natural products.[1]

Over the course of human history, we have acknowledged *Capsicum* not only as a spice, but as a medicinal plant. In pre-Hispanic societies, healers used preparations with chilli peppers to treat respiratory issues, stomach ache, pain and inflamed injures. Yet, as with any medicine, *Capsicum* must be used correctly. Father Acosta first described the pharmacodynamics of chilli peppers in the sixteenth century. He observed that the moderate ingestion of ají chilli peppers could aid digestive disorders, whereas excessive consumption could be harmful, causing gastritis.[2]

The medicinal properties of *Capsicum* are closely related to its chemical composition. As we have seen in earlier chapters, the chemical profile of *Capsicum* is, in general, similar to other vascular species (higher plants) and consists of classes such as alkaloids, terpenoids, fatty acids and phenolic compounds. The biological properties of such compounds in the treatment of many diseases are well known[3] and they surely also contribute to the biological

Food Chemistry, Function and Analysis No. 37
Chemistry and Nutritional Effects of Capsicum
By Valdir Florencio da Veiga, Jr, Larissa Silveira Moreira Wiedemann, Claudio Pereira de Araujo, Jr and Ananda da Silva Antonio

Published by the Royal Society of Chemistry, www.rsc.org

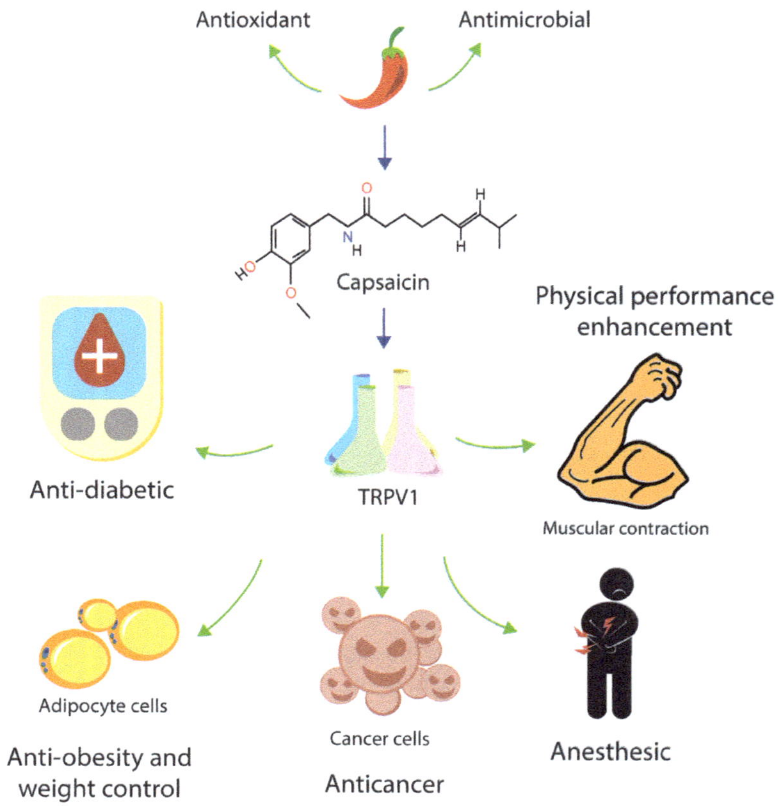

Figure 7.1 Overview of the pharmacological properties of *Capsicum*. Image credit: Carlos Magno da Silva Antonio.

properties of *Capsicum*. However, we must remember that the chemical classes of capsaicinoids and capsinoids are exclusively found in *Capsicum* species. This characteristic of their chemical composition and the unique way in which these compounds interact with the neural receptors of mammals mean that *Capsicum* peppers present very specific biological properties (Figure 7.1).

Among their many biological properties, chillies and sweet peppers are mainly acknowledged for their anticancer, antidiabetic, anti-obesity and neuropathic anaesthetic properties. In addition, *Capsicum* has recently been considered as a thermogenic agent with useful properties that can enhance the physical performance of athletes. This chapter describes how *Capsicum* metabolites interact with the human body to produce these therapeutic properties.

7.2 Pharmacokinetics and Pharmacodynamics

To understand the pharmaceutical implications and potential of specific metabolites or natural products, we need to consider two major concepts:

the pharmacokinetics and the pharmacodynamics. Pharmacokinetics refers to how organs process a drug – in other words, how the organism absorbs, distributes and metabolises it. Pharmacodynamics refers to the effects of the drug on the organism, or the interaction of the drug with any structure within the body, such as a neural receptor. Both processes are extremely important in optimising the biological activity of an intended medicine, to avoid toxic effects and to ensure that the dose used produces the desired medicinal effects.

Despite the presence of several active metabolites in *Capsicum*, capsaicin is always an interesting example because it is only be found in this single genus. When considering the pharmacokinetics of capsaicin, the first question to answer is, "how can it be administrated?". As capsaicin is poorly soluble in water, topical and oral administration are preferred.

Topical administration is usually performed with preparations containing 0.025–0.075% capsaicin, although preparations can contain higher concentrations (up to 10%). The vehicle in such preparations is usually 70% isopropyl alcohol, 70% mineral oil or propylene glycol. Within such preparations, capsaicin can be detected in the stratum corneum within 1 min of application, reaching its maximum concentration within 8–15 minutes, depending on the drug vehicle.[4,5] The topical application of capsaicin preparations has a delivery rate of nearly 2.71 $\mu g\,cm^{-2}\,h^{-1}$ to the systemic circulation, reaching a maximum concentration in the stratum corneum and plasma of 16 μg and 17.8 $ng\,mL^{-1}$, respectively, even when using preparations with concentrations as high as 30% capsaicin.[4,6]

Oral administration allows capsaicin to be absorbed in the range 50–94% within 1 h of ingestion.[7] Absorption by oral ingestion occurs quickly, mainly in the stomach and intestines *via* passive transport (diffusion and facilitated diffusion) with no energy expenditure or *via* active transport proteins (*e.g.* albumin). The capsaicinoids are broadly distributed through the organism, although the liver and kidneys will concentrate the greatest amounts (nearly 25% of the total administrated dose) of capsaicinoids in the tissues within 24 h of administration. Twenty-four hours after ingestion, the capsaicinoid concentration in the liver and kidneys will have decreased by nearly 90%.[8] Thus, despite the fact that capsaicinoids diffuse to several organs, the human body metabolises them quickly, decreasing the chance of overdose or subacute and chronic intoxication.

The absorption and metabolism of capsaicinoids vary depending on the route of administration. The biotransformation of capsaicin is slower after dermal application. Although other tissues (*e.g.* the intestines, liver and stomach) can biotransform an average of 90% of orally ingested capsaicin, the skin is only able to metabolise an average of 75% with a half-life of about 2 h. The total metabolism time for capsaicin applied topically is >20 h and this is the preferred route of administration if long exposure is required.[9,10]

The route of administration influences the pharmacokinetics of capsaicinoids. Several different technologies can be used during the formulation of *Capsicum* bioproducts to optimise their delivery and biological interaction

Table 7.1 Pharmacokinetic parameters of capsaicin in blood samples after oral administration.[a]

Pharmaceutical form	Dose ($mg\,kg^{-1}$)	Model	Sample	C_{max} ($ng\,mL^{-1}$)	T_{max} (h)	$T_{1/2}$ (h)	Ref.
Capsaicin	1.5	Rat	Plasma	392.07	0.25	5.19	11
Capsaicin self-assembled nanoparticle	1.5	Rat	Plasma	751.84	1	8.01	11
Capsaicin	30	Rat	Serum	1.90	1	6	12
Capsaicin	60	Rabbit	Plasma	262.62	1	4.29	13
Solid dispersion of capsaicin	60	Rabbit	Plasma	384.30	1	5.9	13
Matrix pellets of capsaicin	60	Rabbit	Plasma	492.06	3	9.40	13
Capsaicin	90	Rat	Plasma	808	0.75	4.61	14
Capsaicin-loaded micelles	90	Rat	Plasma	702	4	19.4	14
Capsaicin-loaded liposomes	90	Rat	Plasma	846	4	29.1	8

[a] C_{max} = maximum concentration; T_{max} = time required to reach maximum concentration; $T_{1/2}$ = half-life.

with an organism (Table 7.1). Such technologies focus on controlling the delivery of capsaicinoids, even after consumption. For instance, instead of absorption starting in the mouth, we can modify the delivery mechanism so that absorption only begins in the intestines at a specific concentration over a desired period of time. We need to control the release of drugs based on the time, duration and/or location of absorption to produce more technological, versatile and effective medicinal bioproducts.

Among the various technologies used to increase the bioavailability of capsaicinoids, encapsulation within micelles and liposomes can be highlighted.[8,11] The enhancement of bioavailability by liposome encapsulation (Table 7.1) occurs because liposomes act as a reservoir system, modulating the release of capsaicin over a longer time period in specific tissues that have similar properties to the liposome coating. Zhu *et al.*[8] showed that the application of encapsulation technology also decreased intestinal irritation in patients consuming capsaicin formulations. In cancer treatment, capsaicin has been encapsulated in a liposome composed of glycerophospholipids that only release capsaicin after interaction with the phospholipase-2 enzyme, which is commonly expressed in cancer cells.[7]

After ingestion and transportation to the systematic circulation, capsaicin is mainly metabolised by hepatic enzymes in the liver to 16-hydroxycapsaicin, 17-hydroxycapsaicin and 16,17-dihydrocapsaicin (Figure 7.2). These metabolites have no pharmacological activity and are readily eliminated *via* the renal pathway, mainly in urine. Other biotransformation products that can be eliminated include vanillic alcohol, vanillic acid and vanillylamine. The non-metabolised fraction of capsaicin (6–10%) is eliminated in the faeces.[12] In skin, most of the capsaicin is metabolised to vanillylamine and vanillic acid,

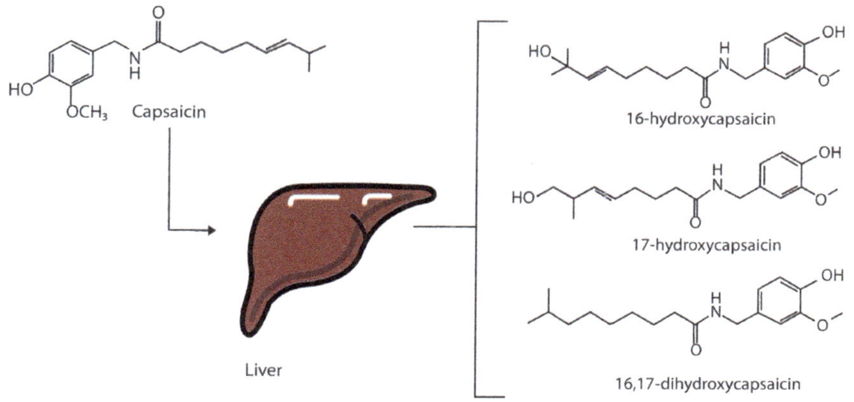

Figure 7.2 Biotransformation products of capsaicin produced by the liver. Image credit: Carlos Magno da Silva Antonio.

confirming that most of the biotransformation of capsaicin occurs within the hepatic system by interaction with cytochrome P450 enzymes.[12]

The cytochrome P450 enzymes are primarily responsible for metabolising capsaicinoids within mammals. This superfamily of enzymes is known for its role in oxidising steroids, fatty acids and xenobiotic compounds *via* hydroxylation (oxygenation). Such interactions produce a series of compounds from capsaicin because the P450 enzymes act on the alkyl chain, the amide group and the aromatic ring in its structure (Figure 7.3).[13]

Capsaicin metabolism by P450 enzymes starts with modifications in the alkyl, amide or aromatic moieties (Figure 7.3). Starting with the alkyl moiety, capsaicin can undergo cyclisation to form a macrocyclic molecule with an amide group and the tertiary allelic carbocation at the end, followed by dehydrogenation or hydrogenation to form alcohols. The amide moiety can also undergo *N*-dehydrogenation to produce a dehydrogenated imide and a further derivative after epoxidation of the aromatic ring. The aromatic moiety can undergo either *O*-demethylation or hydrogenation during capsaicinoid metabolism.[13] The pharmacophore of capsaicinoids includes not only the vanilloid moiety, but also the alkyl chain. Any modification within this structure affects the interactions of capsaicinoids with TRPV1 and therefore even relatively small modifications, such as dehydrogenation in the alkyl chain (Figure 7.3), can limit the pharmacological effects and contribute to the toxicity of this chemical class.[13]

Metabolism by P450 will favour the formation of a specific degradation product depending on the type of capsaicinoid. Capsaicinoids with longer alkyl chains favour the alkyl metabolism pathway (Figure 7.3), leading to the formation of alcohol. Increases in the hydrophobicity of the substrate therefore enhance the enzymatic partitioning of the alkyl chains or allow better penetration of the substrate within the active site of P450, increasing the interaction of the substrate terminus and the iron–oxo haeme.[13,14]

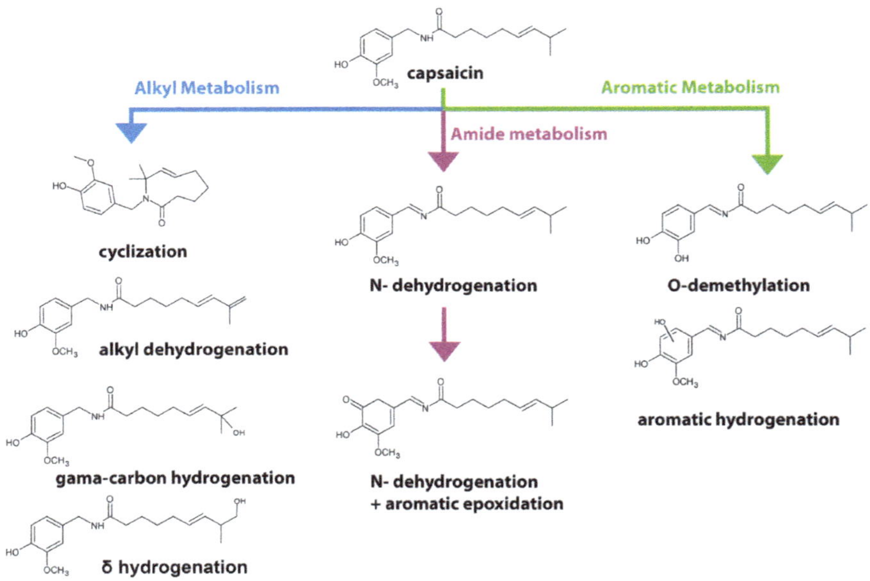

Figure 7.3 Degradation products produced by the interaction of capsaicin with P450 enzymes. Image credit: Carlos Magno da Silva Antonio.

However, alkyl chains more than ten carbon atoms long may inhibit the formation of alcohols during metabolism. The dehydrogenation of the alkyl chain is also influenced by the vanillyl moiety. If the vanillyl moiety undergoes elimination or alkylation at the 4-OH position, then the ability of P450 to dehydrogenate the capsaicinoids is affected because small changes in the vanillyl moiety change the conformation between the enzyme and substrate.

7.3 Capsaicin and the TRPV1 Receptor

The major receptor responsible for the pharmacological and sensorial effects of capsaicinoids is the transient receptor potential cation channel subfamily V member 1 (TRPV1). The interaction of capsaicinoids with TRPV1 is the major pharmacodynamic characteristic of the *Capsicum* species and is currently so well accepted that this receptor is also known as the capsaicin receptor, even though it can interact with any substrate containing a vanilloid moiety. There are several TRPV1 agonists in addition to the capsaicinoids, such as resiniferatoxin, piperine, oleic acid, olvanil, the endovanilloids (anandamide, 12-HPETE, *N*-arachidonoyl dopamine and *N*-oleoyldopamine), endocannabinoid compounds and arachidonic derivatives.[15] TRPV1 is highly expressed on the C and Aδ nerve fibres and is distributed within the brain, sensorial nerves, dorsal root ganglia, bladder, bowel and blood vessels.

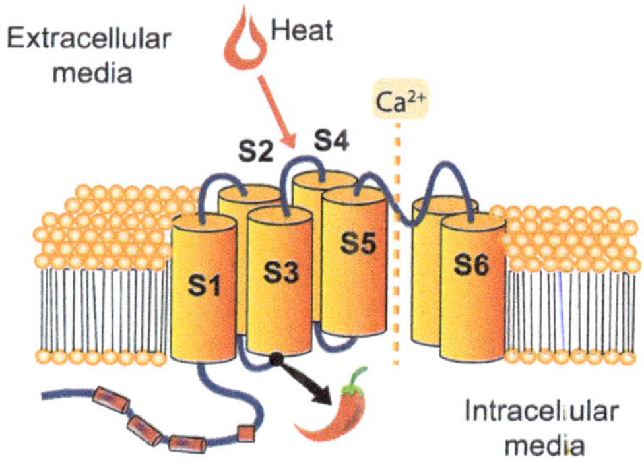

Figure 7.4 Representation of the position of TRPV1 within cellular membranes. Image credit: Carlos Magno da Silva Antonio.

TRPV1 is a somatosensorial cell receptor responsible for the perception of heat. It can be activated by high temperatures (43–45 °C), acidosis (pH < 5.4), tissue damage and inflammatory processes. Its activation is therefore closely related to mechanisms associated with the perception of pain and triggering of the inflammatory response in the body.[15,16] TRPV1 is composed of six transmembrane helix protein Sections with N- and C-termini in the intracellular media (cytosolic side) and a re-entrance loop between Sections S5 and S6 (Figure 7.4). The interaction site between capsaicin and TRPV1 is located within Sections S3, S4 and S4–S5. The capsaicin vanilloid moiety and amide functional group form hydrogen bonds with the protein residues along this site. The activation of TRPV1 by capsaicin is mainly related to the conformational changes that occur as a result of the interaction of the vanilloid moiety with specific residues of the Section S4–S5 link.[17]

Activation of TRPV1 leads to the rapid influx of cationic ions, which depolarise the nociceptive neurons, activating the sting sensation of pungency. Subsequently, its activation modulates the balance of organelle physiology, mainly the mitochondria. After the initial excitation, TRPV1 enters a refractory state, known as desensitisation, responsible for the anaesthetic effects of capsaicin.[18]

7.4 Toxicity and Side-effects

Despite the constant consumption of *Capsicum* worldwide and its remarkable pharmaceutical properties, anything that is consumed in excess can present toxic effects. If you think of the pungency of a hot pepper and the pain associated with it, it is not hard to imagine that there is a certain point at which it can produce severe pain or side-effects. In fact, as we see later in this book, *Capsicum* species can even be used in deterrent devices.

Acute intoxication with capsaicinoids can be caused by exposure to high doses or continuous consumption. In both cases, the constant or intense interaction of capsaicinoids with TRPV1 can cause it to desensitise, along with several other cutaneous nociceptors, disabling the perception of several sensorial features, such as heat sensation or touch. The loss of the touch sensation also involves the anaesthetic effect of capsaicinoids, a property exploited in some painkiller medication.[19]

In addition to desensitisation, overexposure to capsaicinoids can cause mitochondrial dysfunction and inhibition of the Ca^{2+} electron transport chain. Ca^{2+} has a crucial role in signalling pathways activating/inhibiting cellular processes and communication between cells and nerves. The inhibition of Ca^{2+} transport affects the peripheral fibres, inducing cutaneous innervation and making them less likely to be activated by peripheral stimuli. Because these fibres can no longer be activated, the central desensitisation is attenuated. The extended exposure of the receptors to the action of capsaicin causes several effects, including the temporary loss of the membrane potential, an inability to transport neurotrophic factors and the reversible retraction of epidermal nociceptive fibre terminals, causing a block in the transmission of nociceptive stimuli for a prolonged period. However, this condition is temporary and reversible within three months.[20]

For oral and intravenous administration, intoxication with capsaicinoids can induce cellular breakdown; however, this effect has been used to treat cancer and other tumours. If you are thinking, "Is it possible to die eating peppers?", the answer is that an instantaneous fatal effect is very unlikely when eating *Capsicum*. A healthy 70 kg adult would need to ingest nearly 35 mg of pure capsaicin for a lethal effect.[21] However, the excessive ingestion of extremely hot peppers, such as Carolina Reaper, or extensive exposure to pepper sprays can induce cardiovascular and breathing complications in susceptible people with pre-existing conditions.[22]

It is important to understand that the capsaicinoids in natural products are safe. For instance, it has been reported that the high consumption of chilli peppers in Mexico and India (25–200 mg per day) is correlated with a higher incidence of gastrointestinal cancers. However, this correlation was not related to the capsaicinoids in the chilli peppers, but to contamination of the fruits with aflatoxins.[23–25] Aflatoxins are natural toxins produced by moulds that infest agricultural crops. These toxins are carcinogens and can cause immunodeficiency and immunosuppression after chronic exposure. Contamination with aflatoxins is mainly related to poor practices in farming, harvesting, post-harvesting and storage that leave fruits exposed to high levels of moisture and high temperatures. As much of the cultivation of *Capsicum* is performed by middle- to low-income producers with little implementation of industrial resources and quality standards, several cases of the contamination of *Capsicum* bioproducts with aflatoxins have been reported.[26–29]

Early studies of capsaicinoid toxicity showed that the route of administration changes the lethal dose. For instance, the oral and topical administration of capsaicin have a 50% lethal dose (LD_{50}) in mice of 190 and 500 $mg\,kg^{-1}$, respectively, whereas intravenous and intratracheal administration produce LD_{50} values of 0.56 and 1.6 $mg\,kg^{-1}$, respectively.[30]

The mechanism of lethality in such cases is cardiovascular and pulmonary dysfunction and failure. Despite this, the routine use of capsaicinoid-based products is considered to be safe for humans under normal conditions *via* both the topical and oral routes and capsaicinoids are currently classified as GRAS (generally considered safe) substances by the US Food and Drug Administration and are approved as food additives or topical analgesics without an extensive toxicological profile. The toxic effects of capsaicinoids can also be beneficial to the human body. In specific doses, capsaicinoids can reduce tumour growth and induce apoptosis of cancers *in vivo*.[31]

7.5 Pharmaceutical Applications

In addition to their well-known culinary uses, which we will explore in Chapter 11, the biological properties of capsaicinoids are also readily applied to therapeutic products, such as the auxiliary treatment of chronic pain, diabetes and its complications, cancers, tumours and other neoplasms, rheumatism, flu with cough, and headaches.[32,33]

7.5.1 Pain Relief

Capsaicin can be used in the treatment of neuropathic pain *via* external application. It must be made clear that, according to the US Food and Drug Administration, *Capsicum*-derived extracts are not considered safe and effective when used as the active ingredient in non-prescription drugs, particularly those taken orally as digestive aids, anti-thermal treatments or pain relief.[23]

In the treatment of neuropathic pain, capsaicin is usually impregnated in dermal patches (Figure 7.5) at concentrations ranging from 0.025 to 0.1% in non-prescription drugs.[34] The efficacy of capsaicin for such treatments is related to its interaction with TRPV-1. Continuous exposure to capsaicin at low concentrations causes desensitisation of TRPV-1, which promotes the anaesthetic and pain relief effects.[34,35]

Some patch formulations with 8% capsaicin are recommended for HIV-associated neuropathy, post-herpetic neuralgia and painful diabetic peripheral neuropathy. There are some issues with these 8% formulations, such as an initial burning sensation and an increased chance of adverse effects at the application site when compared with low-concentration formulations. Other adverse effects include skin dryness, erythema, oedema, pain, papules and pruritus.[9,36]

7.5.2 Obesity and Diabetes

Obesity is the excessive accumulation of fat in a manner that impairs health. This comorbidity can be caused by a sedentary lifestyle, genetic susceptibility, cultural influences and unrestricted access to an ample supply of high-calorie foods. The homeostatic mechanisms that control caloric

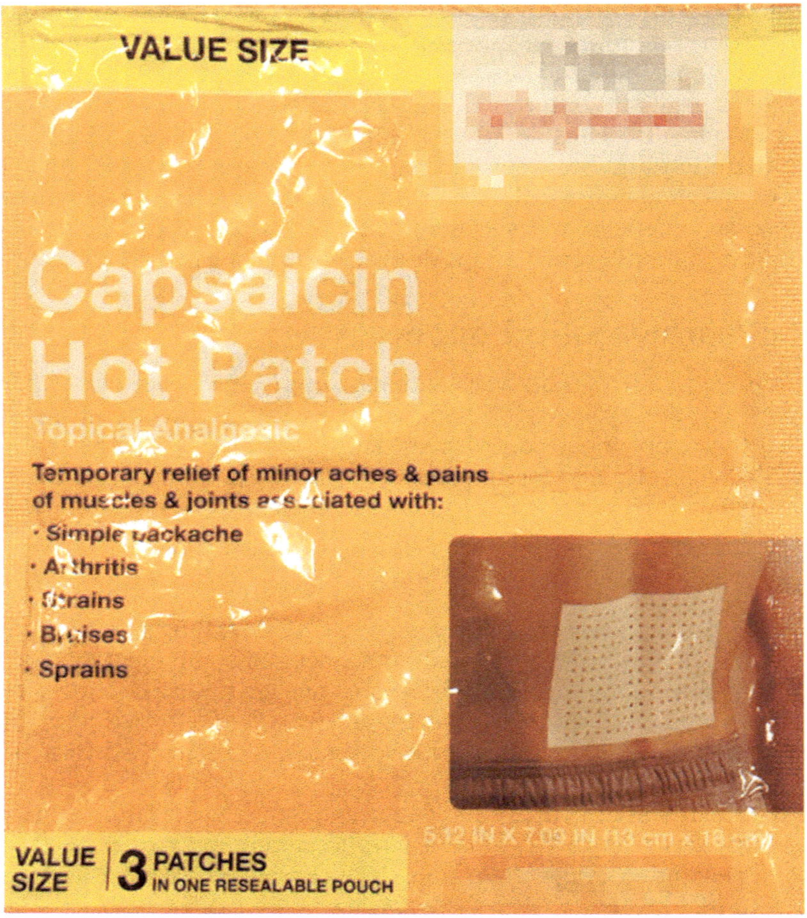

Figure 7.5 Capsaicin patch for the treatment of neuropathic pain. Image credit: Carlos Magno da Silva Antonio.

balance are considerably changed. Obesity alone has a low mortality rate, but the excessive accumulation of fat can cause hypertension, hypercholesterolaemia, type 2 diabetes, cardiovascular disease, strokes, cancers, respiratory diseases and digestive problems, which can lead to death. Nearly one-quarter of the global population is currently obese, emphasising the great public health issue that this comorbidity has become.[37] Although obesity can be reversed by following a healthier lifestyle, some people require pharmacotherapeutic intervention with appetite suppressants (usually sibutramine and orlistat). Such drugs inhibit the uptake of 5-HT and norepinephrine in the hypothalamic region responsible for the regulation of food intake, increasing the sensation of satiety (sibutramine). Appetite suppressants also inhibit lipid degradation and the absorption of lipids into tissues.[38]

The application of *Capsicum* metabolites in the treatment of obesity is based on their thermogenic properties and their ability to increase satiety, suppress hunger, increase energy expenditure and increase fat oxidation.[39] The effect of capsaicinoids on weight loss is related to the activation of TRPV1 (Figure 7.6). As capsaicinoids stimulate TRPV1, this promotes the release of norepinephrine in the sympathetic nervous system, initiating the thermogenesis of fat tissue and upregulating the action of uncoupling protein 1 (UCP1). UCP1 produces energy by oxidising fatty acids and glucose in brown adipose tissue, promoting weight loss.[40,41]

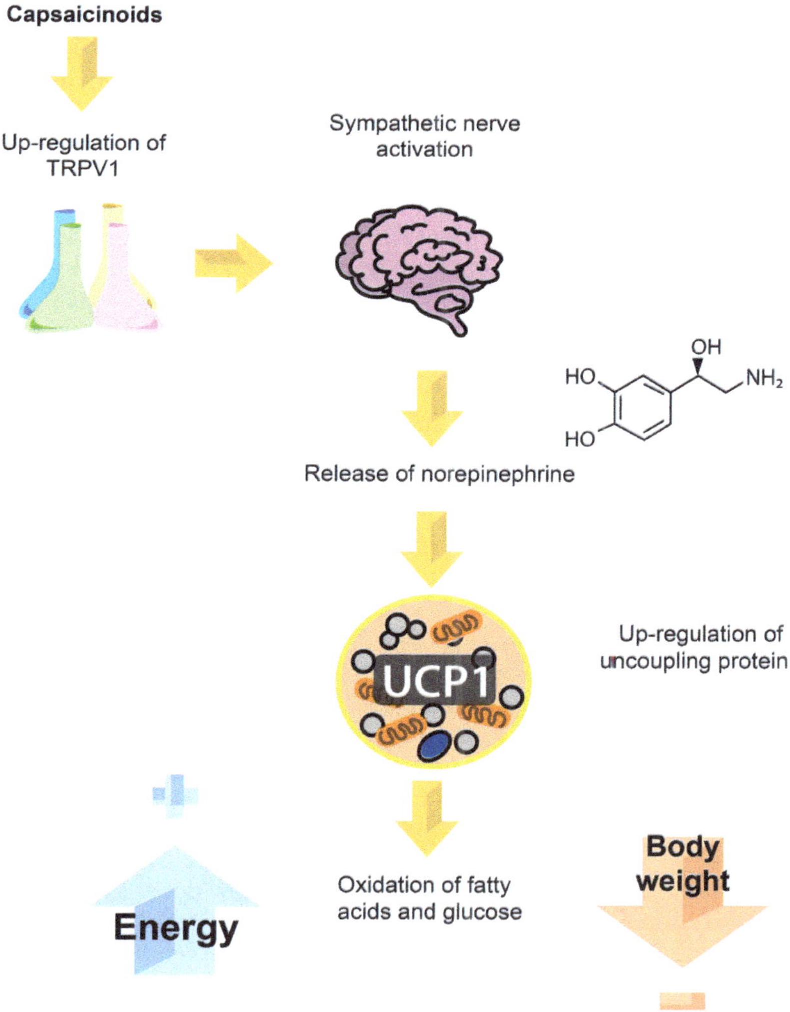

Figure 7.6 Major mechanism of the control of body weight by capsaicinoids. Image credit: Carlos Magno da Silva Antonio.

Another comorbidity that is spreading rapidly is diabetes. It is estimated that 422 million people worldwide have diabetes. There are two types of diabetes: type 1 diabetes, characterised by a lack of insulin; and type 2 diabetes, characterised by the ineffective use of insulin or insulin resistance. Resistance to insulin in skeletal muscles and the liver usually progresses to pancreatic β-cell insufficiency. This resistance unbalances the metabolism of glucose, inducing hyperglycaemia. There is currently no cure for diabetes, only supportive treatments to control insulin levels. Such treatments include the administration of insulin, sulphonylureas, glinides and thiazolidinediones. However, these drugs are also associated with an increased risk of bone fracture and congestive heart failure. These treatments also often lose efficacy within four years, requiring the constant development of new treatments.[32,42,43]

Capsaicinoids can help in the treatment of diabetes because they reduce blood glucose levels and the resistance to insulin. The mechanisms are complex and not yet completely elucidated. The influx of Ca^{2+} is significantly increased during the activation of TRPV1 by capsaicinoids because TRPV1 is a Ca^{2+}-exchange channel. Stimulation of the influx of Ca^{2+} enhances the secretion of glucagon-like peptide 1 (GLP-1) by the pancreatic β-cells. The major function of GLP-1 is to increase the production of insulin, which enables the absorption of glucose from the blood and reduces hyperglycaemia.[44,45]

Another mechanism induced by capsaicinoids that aids the control of diabetes is the activation of 5′-adenosine monophosphate-activated protein kinase (AMPK). AMPK is an enzyme that stimulates the absorption of glucose by the skeletal muscles and its activation also induces lipolysis and lipid metabolism in the liver and muscles. Because AMPK can promote the absorption of glucose *via* an insulin-independent mechanism, it can reduce insulin resistance in people with type 2 diabetes.[41,46]

7.5.3 Cancer

Cancer is a large group of diseases that can affect any tissue and is characterised by the uncontrolled growth of abnormal cells to the point at which the dysfunctional cells completely invade the original tissue and other organs (metastasis), eventually causing organ failure and death. Cancer treatment is based on radiation therapies, chemotherapy and drug administration. A key factor in metastasis is angiogenesis, in which new blood vessels are created, enabling the tissue to grow. Cancer cells proliferate in different organs and create tumours *via* angiogenesis. The inhibition of angiogenesis is therefore one of the strategies to slow down cancer and tumour progression.[47]

Capsaicinoids and capsinoids can inhibit vascular endothelial growth factor (VEGF), one of the major proteins inducing the growth of blood vessels.[48,49] These inhibitory properties are independent of the interaction of capsaicinoids with TRPV1, although the exact mechanism of the inhibition of angiogenesis is still unclear.[50] The anticancer properties of *Capsicum* can also be observed in other processes involved in the development of cancer, such as apoptosis.[31]

7.5.4 Exercise Performance

Most people do not know that *Capsicum* can improve exercise performance. To perform any exercise that requires a high energy expenditure with few or no rest periods, such as short- and middle-distance running, the muscles need to be prepared to contract and release energy. This kind of exercise is known as anaerobic exercise, in which a large amount of energy must be released by constant contractions of the skeletal muscles in a period shorter than that necessary to supply oxygen to the body.[51]

Contraction of the skeletal muscles occurs *via* contraction of the sarcomere within muscular fibres. The sarcomere is composed of filaments of actin and myosin proteins. To contract, myosin and actin increase their binding with each other, following a "sliding" movement, shortening the sarcomere and, consequently, causing muscular contraction (Figure 7.7). For this reaction to occur, there must be an influx of Ca^{2+} and ATP into the sarcomere to produce the shorter complex between actin and myosin. As capsaicin stimulates TRVP1 and is a Ca^{2+} channel, their interaction within skeletal muscles increases the release of Ca^{2+}, favouring the reaction between actin and myosin. This makes muscular contraction easier and enhances the production of force.[52]

The increased release of Ca^{2+} promoted by capsaicin also enhances the efficacy of intramuscular triglyceride depletion. This process optimises the consumption of adipocere tissue (lipolysis) and triglycerides during the resynthesis of ATP in continuous muscle contractions and preserves muscle glycogen. Capsaicin is therefore considered to be a ergogenic agent, increasing lipolysis, reducing lipogenesis and enhancing the production of force.[53]

Studies conducted in mouse models using a four-week daily administration of capsaicin showed that capsaicin supplementation reduces serum lactate, ammonia, blood urea nitrogen and creatine kinase concentrations in blood after intense and short-term duration exercise, which are all indicators of muscle fatigue.[54] By contrast, most experiments in humans have shown that blood lactate concentrations are not affected by capsaicin supplementation, although performance (*e.g.* the running time across a specific distance) was improved.[55] The decrease in the perception of exertion is suggested to be partially caused by the analgesic effects of capsaicinoids. The evaluation of capsaicin supplementation to increase physical performance is usually based on the oral administration of 12 mg of pure capsaicin, a dose considered tolerable to humans without gastrointestinal discomfort.[55–57]

7.5.5 Antimicrobial Activity

With the increasing resistance of bacteria and other microorganisms to drug treatments, many research groups are exploring natural products to develop

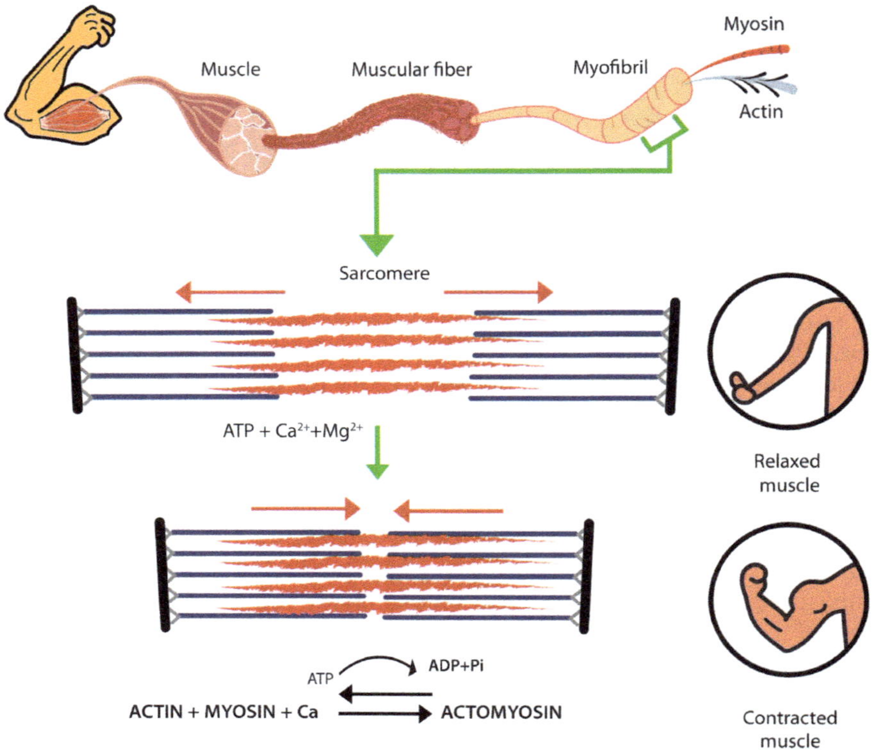

Figure 7.7 Schematic illustration of the muscle contraction system. Image credit: Carlos Magno da Silva Antonio.

new antibiotic agents.[58] Several reports have described the potential antimicrobial activity of different extracts from *Capsicum* (Table 7.2).[39] Many metabolites found in the *Capsicum* genus have antimicrobial activity, such as flavonoids, terpenoids and peptides.

The antimicrobial activity of *Capsicum* bioproducts (Table 7.2) is of great use in food additives, increasing the shelf life of products such as raw meat, sauces and condiments. Another promising antimicrobial activity attributed to capsaicin is against *Helicobacter pylori*. *H. pylori* is a gram-negative bacterium that affects >50% of world's population, causing severe gastritis, peptic ulcers, adenocarcinomas and lymphoma of the associated gastric mucosal lymphoid tissue. Even with so many people infected, currently no single antibiotic treatment is capable of eradicating this microorganism. Treatment is based on a combination of antibiotic and antisecretory agents, such as proton pump inhibitors or histamine receptor antagonists. However, such treatments can cause side-effects ranging from gastric disorders to resistance of *H. pylori* against the antibiotics. The defensive mechanisms of *H. pylori* have led to a continuous search for new antibiotics from synthetic or natural sources.

Table 7.2 Antimicrobial activity of *Capsicum* preparations.[a]

Sample tested	Bacteria species	MIC (%)	Ref.
Capsicum nanoemulsion oleoresin	*Staphylococcus aureus*	NA	61
	Escherichia coli	NA	
Capsicum baccatum oleoresin	*Bacillus cereus*	0.25	62
Capsicum chinense oleoresin	*Bacillus cereus*	0.25	62
Capsicum frutescens oleoresin	*Staphylococcus aureus*	0.25	62
	Aspergillus niger	0.10	
	Penicilium sp.	0.25	
Capsicum frutescens ethanol extract	*Proteus vulgaris*	NA	63
	Pseudomonas aeruginosa	NA	
	Escherichia coli	NA	
	Bacillus subtilis	NA	
	Bacillus megaterium	NA	
	Staphylococcus aureus	NA	
	Candida albicans	NA	
Capsicum annuum jalapeño	*Listeria monocytogenes*	NA	64
	Escherichia coli	NA	
	Salmonella anatum	NA	

[a]MIC = minimum inhibitory concentration; NA = data not available.

In the last decade, capsaicin has proved to be effective in inhibiting metronidazole-resistant and metronidazole-sensitive *H. pylori* at a minimum concentration of 25 $\mu g\, mL^{-1}$. Its inhibitory effects against *H. pylori* are related to the inhibitory properties of capsaicin against inflammatory factors such as TNF-α. Other mechanisms that aid capsaicin activity against *H. pylori* include its inhibitory effect on urease, an enzyme essential for the survival and pathogenicity of *H. pylori*. The concentration required to achieve such an effect varies from 31.25 to 1000 $\mu g\, mL^{-1}$. Although some people think that *Capsicum* should not be consumed by people with gastritis or ulcers, the consumption of small dietary amounts of *Capsicum* may be beneficial in diseases caused by *H. pylori* infection. The synergistic effect of capsaicin with metronidazole can aid the effectiveness of treatment.[59,60]

7.5.6 Antioxidant and Anti-inflammatory Activities

One of the most studied biological activities of natural products is their antioxidant activity. This property is functional in human health as it describes the ability to limit, restrain or inhibit oxidative processes, including those responsible for ageing or abiotic stress (*e.g.* exposure to solar radiation). Antioxidants have a huge role as protective agents against many diseases and health conditions, such as cancer, diabetes, cardiovascular disease and anaemia.[61] Several methods are used to evaluate the antioxidant potential of a bioproduct and these are mainly divided in two categories.[61]

1. Direct methods measure the ability to inhibit lipid oxidation, such as the ORAC assay, the β-carotene linoleic acid method and luminol chemiluminescence.
2. Indirect methods measure the ability to eliminate free radicals, such as the DPPH (1,1-diphenyl-2-picrylhydrazyl) scavenging test, the ABTS (2,2-azino-bis(3-ethylbenzthiazoline-6-sulfonic acid)) radical cation decolourisation assay and the Trolox-equivalent antioxidant test.

Several classes of chemicals present antioxidative properties, many of which are found in *Capsicum*, such as flavonoids, carotenoids, fatty acids, phenolic acids, alkaloids and the capsaicinoids. The antioxidant capacity will be different depending on the qualitative and quantitative profile of each one of these classes within a species or its bioproducts (Table 7.3).

Any factor that influences the chemical composition of a bioproduct will affect the antioxidant activity. For instance, because the maturity stage of *Capsicum* modifies the carotenoid composition, it also modifies the antioxidant capacity of the fruit. Because each species, variety and cultivar has a specific chemical profile, their antioxidant capacities will also be different (Table 7.3). In *Capsicum*, it is usually assumed that capsaicinoids and the phenolic content have the strongest correlation with antioxidant tests, mainly using indirect measuring methods.[62–64] For direct methods, the antioxidant properties of *Capsicum* are mainly attributed to the capsaicinoids because they can reduce metals, such as iron and copper, which mediate the peroxidation of low-density lipoproteins.[65]

The capsaicinoids also present anti-inflammatory activity because they can mediate the expression of the prostaglandins, cyclooxygenase-2 (COX-2) and nitric oxide synthase (iNOS), which are key players in the pathogenesis

Table 7.3 DPPH (1,1-diphenyl-2-picrylhydrazyl) scavenging test results for *Capsicum* bioproducts.

Capsicum species	DPPH test	Units for measuring inhibition	Ref.
C. annuum yellow cayenne fresh	359.9	IC_{50} μg mL^{-1}	62
C. baccatum ají angelo fresh	468.8	IC_{50} μg mL^{-1}	62
C. chinense capezzolo di scimmia fresh	89.5	IC_{50} μg mL^{-1}	62
C. frutescens shimatogarashi seeds (green maturity stage)	4	μmol Trolox-equivalent (g fresh weight)$^{-1}$	63
C. frutescens shimatogarashi seeds (orange maturity stage)	5	μmol Trolox-equivalent (g fresh weight)$^{-1}$	63
C. frutescens shimatogarashi seeds (red maturity stage)	5.9	μmol Trolox-equivalent (g fresh weight)$^{-1}$	63
C. annuum jalapeño hexane extract	95.1	% of inhibition	64
C. annuum serrano hexane extract	92.4	% of inhibition	64
C. annuum jalapeño methanol extract	45.8	% of inhibition	64
C. annuum serrano methanol extract	59.3	% of inhibition	64
C. chinense biquinho	18.04	IC_{50} μg mL^{-1}	65

of inflammation. Most common non-steroidal anti-inflammatory drugs, such as aspirin, inhibit COX-2 activity, which also reduces the production of prostaglandins and nitric oxide radicals. The anti-inflammatory properties of capsaicin *in vitro* are based on the modulation of the expression/catalytic activity of COX-2, which can induce apoptosis and control tumour proliferation.[66,67]

7.6 Concluding Remarks

Capsicum fruits are famous worldwide for their pungency, but few people know that the same biological mechanism that promotes pungency also confers bioactivity in these species. The chemical profile of this genus and the unique presence of capsaicinoids enables their use as anaesthetic, antioxidant, antimicrobial, antidiabetic, antihypertensive and anticarcinogenic agents. Pharmacological applications of *Capsicum* are cheaper and have fewer side-effects than synthetic drugs and can therefore be used to improve public health. However, we need to fully comprehend the interactions of *Capsicum* metabolites with the human body, its pharmacophore, pharmacodynamics and pharmacokinetics before further exploitation of this genus in medical treatments.

Acknowledgements

The authors acknowledge the precious help of Ian Gardel Carvalho Barcellos Silva.

References

1. Z. Wang and L. Yang, *Front. Pharmacol.*, 2020, **11**, 1–12.
2. M. J. Hass, University of Montana, 1994.
3. N. Joshee, S. A. Dhekney and P. Parajuli, *Medicinal plants: From farm to pharmacy*, Springer Nature Switzerland, Cham, 2019.
4. L. K. Pershing, C. A. Reilly, J. L. Corlett and D. J. Crouch, *Toxicol. Appl. Pharmacol.*, 2004, **200**, 73–81.
5. M. Hayman and P. C. A. Kam, *Curr. Anaesth. Crit. Care*, 2008, **19**, 338–343.
6. A. D. P. Papoiu and G. Yosipovitch, *Expert Opin. Pharmacother.*, 2010, **11**, 1359–1371.
7. W. D. Rollyson, C. A. Stover, K. C. Brown, H. E. Perry, C. A. M. Cathryn, D. Stevenson, J. G. Ball, M. A. Valentovic, P. Dasgupta, C. D. Stevenson, C. A. McNees, J. G. Ball, M. A. Valentovic and P. Dasgupta, *J. Controlled Release*, 2010, **0**, 96–105.
8. Y. Zhu, M. Wang, J. Zhang and W. Peng, *Arch. Pharmacal Res.*, 2014, **38**, 512.
9. H. A. Blair, *Drugs*, 2018, **78**, 1489–1500.
10. M. Lu, C.-T. Ho and Q. Huang, *J. Food Drug Anal.*, 2017, **25**, 27–36.

11. Y. Zhu, W. Peng, J. Zhang, M. Wang, C. K. Firempong, C. Feng, H. Liu, X. Xu and J. Yu, *J. Funct. Foods*, 2014, **8**, 358–366.
12. D. Suresh and K. Srinivasan, *Indian J. Med. Res.*, 2010, **131**, 682–691.
13. C. a Reilly and G. S. Yost, *Drug Metab. Rev.*, 2006, **38**, 685–706.
14. Y. J. Surh and S. Sup Lee, *Life Sci.*, 1995, **56**, 1845–1855.
15. S. J. Conway, *Chem. Soc. Rev.*, 2008, **37**, 1530–1545.
16. L. Darré and C. Domene, *Mol. Pharmacol.*, 2015, **12**, 4454–4465.
17. F. Yang, X. Xiao, W. Cheng, W. Yang, P. Yu, Z. Song, V. Yarov-Yarovoy and J. Zheng, *Nat. Chem. Biol.*, 2015, **11**, 518–524.
18. F. Yang and J. Zheng, *Protein Cell*, 2017, **8**, 169–177.
19. P. Patowary, M. P. Pathak, K. Zaman, P. S. Raju and P. Chattopadhyay, *Biomed. Pharmacother.*, 2017, **96**, 1501–1512.
20. S. Idrees, M. A. Hanif, M. A. Ayub, A. Hanif and T. M. Ansari, in *Medicinal Plants of South Asia*, ed. M. A. Hanif, H. Nawaz, M. M. Khan and H. J. Byrne, Elsevier, Amsterdam, 2020, pp. 113–124.
21. J. T. Arnold, S. B. Stewart and L. Sammut, *Res. Rev. J. Herb. Sci.*, 2017, **5**, 1–5.
22. S. Toprak, G. Ersoy, J. Hart and P. Clevestig, *J. Forensic Leg. Med.*, 2015, **29**, 36–42.
23. W. Johnson, *Int. J. Toxicol.*, 2007, **26**, 3–106.
24. E. D. van Asselt, J. L. Banach and H. J. van der Fels-Klerx, *Food Control*, 2018, **83**, 7–17.
25. W.-J. Yu, H.-P. Liu, X.-W. Zhang, D. Dong, Y. Jiang, N.-X. Sun, Y.-K. Liu and J.-F. Yuan, *J. Food Process. Preserv.*, 2017, **42**, e13522.
26. S. Z. Iqbal, A. Mumtaz, Z. Mahmood, M. Waqas, A. Ghaffar, A. Ismail and W. Pervaiz, *Food Control*, 2021, **121**, 107621.
27. S. V. Reddy, D. Kiran Mayi, M. Uma Reddy, K. Thirumala-Devi and D. V. R. Reddy, *Food Addit. Contam.*, 2001, **18**, 553–558.
28. L. Zhang, W. Xu, P. Yue, Q. Wang, Y. Li, X. Pei and P. Zeng, *Food Control*, 2020, **110**, 107034.
29. P. Singh and P. J. Cotty, *Int. J. Food Microbiol.*, 2019, **289**, 145–153.
30. P. Piyachaturawat, T. Glinsukon and C. Toskulkao, *Toxicol. Lett.*, 1983, **16**, 351–359.
31. R. Clark and S. H. Lee, *Anticancer Res.*, 2016, **36**, 837–844.
32. K. Singletary, *Nutr. Today*, 2011, **46**, 33–47.
33. B. K. Saleh, A. Omer and B. Teweldemedhin, *MOJ Food Process. Technol.*, 2018, **6**, 325.
34. O. M. Hall, A. Broussard, T. Range, M. A. Carroll Turpin, S. Ellis, V. M. Lim, E. M. Cornett and A. D. Kaye, *Curr. Pain Headache Rep.*, 2020, **24**, 53.
35. H. Knotkova, M. Pappagallo and A. Szallasi, *Clin. J. Pain*, 2008, **24**, 142–154.
36. C. Bonezzi, A. Costantini, G. Cruccu, D. M. M. Fornasari, V. Guardamagna, V. Palmieri, E. Polati, P. Zini and A. H. Dickenson, *Expert Opin. Pharmacother.*, 2020, **21**, 1377–1387.
37. J. Rogers, S. L. Urbina, L. W. Taylor, C. D. Wilborn, M. Purpura, R. Jäger and V. Juturu, *BMC Obes.*, 2018, **5**, 1–10.

38. K. M. M. B. Carvalho, T. S. De Melo, K. M. De Melo, A. L. G. Quinderé, F. T. B. De Oliveira, A. F. S. C. Viana, P. I. G. Nunes, J. D. S. Quetz, D. D. A. Viana, A. A. D. C. A. Da Silva, A. Havt, S. G. D. C. Fonseca, M. H. Chaves, V. S. Rao and F. A. Santos, *Planta Med.*, 2017, **83**, 285–291.
39. M. D. M. Chamikara, D. R. R. P. Dissanayake, M. Ishan and S. D. S. S. Sooriyapathirana, *Ceylon J. Sci.*, 2016, **45**, 5.
40. M. Saito, T. Yoneshiro and M. Matsushita, *Trends Endocrinol. Metab.*, 2015, **26**, 585–587.
41. S. Varghese, P. Kubatka, L. Rodrigo, K. Gazdikova, M. Caprnda, J. Fedotova, A. Zulli, P. Kruzliak and D. Büsselberg, *Int. J. Food Sci. Nutr.*, 2017, **68**, 392–401.
42. I. Tolan, D. Ragoobirsingh and E. Y. S. A. Morrison, *Phytother. Res.*, 2001, **15**, 391–394.
43. D. Y. Kwon, Y. S. Kim, S. Y. Ryu, M. R. Cha, G. H. Yon, H. J. Yang, M. J. Kim, S. Kang and S. Park, *J. Nutr. Biochem.*, 2013, **24**, 1078–1085.
44. S. K. Panchal, E. Bliss and L. Brown, *Nutrients*, 2018, **10**, 14–18.
45. A. Suri and A. Szallasi, *Trends Pharmacol. Sci.*, 2008, **29**, 29–36.
46. S. H. Kim, J. T. Hwang, H. S. Park, D. Y. Kwon and M. S. Kim, *Biochem. Biophys. Res. Commun.*, 2013, **439**, 66–70.
47. J. R. Friedman, S. D. Richbart, J. C. Merritt, H. E. Perry, K. C. Brown, A. T. Akers, N. A. Nolan, C. D. Stevenson, J. D. Hurley, S. L. Miles, M. T. Tirona, M. A. Valentovic and P. Dasgupta, in *Advances in Cancer Research*, ed. P. Fisher and K. Tew, Academic Press, Cambridge, 1st edn, 2019, vol. 144, pp. 263–298.
48. B. J. Pyun, S. Choi, Y. Lee, T. W. Kim, J. K. Min, Y. Kim, B. D. Kim, J. H. Kim, T. Y. Kim, Y. M. Kim and Y. G. Kwon, *Cancer Res.*, 2008, **68**, 227–235.
49. J. K. Min, K. Y. Han, E. C. Kim, Y. M. Kim, S. W. Lee, O. H. Kim, K. W. Kim, Y. S. Gho and Y. G. Kwon, *Cancer Res.*, 2004, **64**, 644–651.
50. L. G. Braga Ferreira, J. V. Faria, J. P. S. dos Santos and R. X. Faria, *Eur. J. Pharmacol.*, 2020, **887**, 173356.
51. C. S. Padilha, F. Billaut, C. Figueiredo, V. L. G. Panissa, F. E. Rossi and F. S. Lira, *Int. J. Sports Med.*, 2020, **41**, 1061–1066.
52. S. L. Volpe, *ACSM's Heal. Fit. J.*, 2020, **24**, 31–32.
53. V. E. L. de Moura e Silva, J. M. Cholewa, F. Billaut, R. Jäger, M. C. de Freitas, F. S. Lira and F. E. Rossi, *Int. J. Sports Physiol. Perform.*, 2020, **16**, 464.
54. Y. J. Hsu, W. C. Huang, C. C. Chiu, Y. L. Liu, W. C. Chiu, C. H. Chiu, Y. S. Chiu and C. C. Huang, *Nutrients*, 2016, **8**, 1–15.
55. M. C. de Freitas, F. Billaut, V. L. G. Panissa, F. E. Rossi, C. Figueiredo, E. C. Caperuto and F. S. Lira, *Eur. J. Appl. Physiol.*, 2019, **119**, 971–979.
56. L. A. Costa, M. C. Freitas, J. M. Cholewa, V. L. G. Panissa, F. Y. Nakamura, V. E. L. M. Silva, A. M. Sá, P. A. Q. Rossi, S. L. G. Ribeiro, M. A. P. Santos, N. E. Zanchi, F. S. Lira and F. E. Rossi, *Int. J. Exercise Sci.*, 2020, **13**, 755–765.

57. M. C. Freitas, J. M. Cholewa, L. A. Gobbo, J. V. N. S. Oliveira, F. S. Lira and F. E. Rossi, *J. Strength Cond. Res*, 2017, **32**, 572–577.
58. B. Singh, *Botanical Leads for Drug Discovery*, Springer, Singapore, 2020.
59. I. Tayseer, T. Aburjai, L. Abu-Qatouseh, N. AL-Karabieh, W. Ahmed and A. Al-Samydai, *J. Pure Appl. Microbiol.*, 2020, **14**, 1.
60. F. Yildiz Zeyrek and E. Oguz, *Ann. Microbiol.*, 2005, **55**, 41.
61. P. L. A. Nascimento, T. C. E. S. Nascimento, N. S. M. Ramos, G. R. Silva, J. E. G. Gomes, R. E. A. Falcão, K. A. Moreira, A. L. F. Porto and T. M. S. Silva, *Molecules*, 2014, **19**, 5434–5447.
62. Manikharda, M. Takahashi, M. Arakaki, K. Yonamine, F. Hashimoto, K. Takara and K. Wada, *J. Oleo Sci.*, 2018, **67**, 113–123.
63. L. S. Santos, C. C. F. L. S. Santos, I. P. B. de Deus, T. L. de Sousa and M. L. Miranda, *Food Sci. Technol.*, 2021, **41**, 497.
64. F. Fratianni, A. D'acierno, A. Cozzolino, P. Spigno, R. Riccardi, F. Raimo, C. Pane, M. Zaccardelli, V. T. Lombardo, M. Tucci, S. Grillo, R. Coppola and F. Nazzaro, *Antioxidants*, 2020, **9**, 1–16.
65. M. R. Loizzo, A. Pugliese, M. Bonesi, F. Menichini and R. Tundis, *LWT–Food Sci. Technol.*, 2015, **64**, 623–631.
66. Z. Ghiasi, F. Esmaeli, M. Aghajani, M. Ghazi-Khansari, M. A. Faramarzi and A. Amani, *Int. J. Pharm.*, 2019, **559**, 341.
67. C. S. Kim, T. Kawada, B. S. Kim, I. S. Han, S. Y. Choe, T. Kurata and R. Yu, *Cell. Signalling*, 2003, **15**, 299.

CHAPTER 8

Capsicum *Oleoresins*

8.1 Introduction

The dictionary definition of an oleoresin is a naturally occurring viscous mixture of resin and essential oils. Oleoresins recall the days of prospectors, who went out into the woods to cut the surface of a tree, before waiting long hours to collect the highly viscous liquid exudate. The term is commonly used in reference to copaiba oil (from the genus *Copaifera* in the Fabaceae family), breu branco oil (from *Protium heptaphyllum*, a tree in the Burseraceae family) and commercial *Pinus* (from the Pinaceae family) species. So how do oleoresins fit into pepper science? There is a second definition for oleoresins, typically related to pharmacological terms, which states that an oleoresin is a viscous mixture obtained from plants *via* extraction with solvents and containing volatile and non-volatile compounds. Oleoresin is a partially an essential oil and partially a semi-solid (Figure 8.1).

In the case of *Capsicum*, the oleoresin is an extract from the dry matter of the fruit, or a part of it, obtained *via* extraction with organic solvents.[1] Depending on the extraction method and the raw material, the oleoresin of *Capsicum* has a density (at 20 °C) of 0.91–0.93 $kg\,m^{-3}$ and a viscosity (at 20 °C) of 43–64 cP.[2] With a characteristic reddish colour, it is an important bioproduct because it contains most of the bioactive compounds found in *Capsicum*, such as capsaicinoids, carotenoids, carbohydrates, flavonoids, vitamins C and E, provitamin A and fatty acids. In fact, oleoresin of *Capsicum* is the major bioproduct of the genus and is used extensively as a food additive or food colouring, an active compound in non-lethal weapons and repellents, and in the manufacture of cosmetics and medicines. *Capsicum* species can be cultivated to present a specific range of specialised metabolites in the maturing fruit for objective use.[3,4] The chemical profile of

Food Chemistry, Function and Analysis No. 37
Chemistry and Nutritional Effects of Capsicum
By Valdir Florencio da Veiga, Jr, Larissa Silveira Moreira Wiedemann,
Claudio Pereira de Araujo, Jr and Ananda da Silva Antonio

Published by the Royal Society of Chemistry, www.rsc.org

Figure 8.1 Paprika oleoresin. Image credit: Carlos Magno da Silva Antonio.

oleoresin of *Capsicum* will be influenced by the species, the cultivation method, storage and the extraction method. Its composition is therefore both varied and complex, creating a diverse array of applications.

8.2 Importance and Applications of Oleoresin of *Capsicum*

The exploitation of natural products has been an important trend in recent decades. Bioproducts have been known throughout human history as a valuable instrument in improving our quality of life. Natural products are not only a source of bioactive compounds and tasty flavours and aromas, but are also less aggressive to our bodies than synthetic formulations.[5] Their use in medicine and cosmetics can therefore enhance the health benefits of these products.

The sensory attributes and biological activities of the compounds present in *Capsicum* fruits are the main reasons for their worldwide use and commercial applications. The current and growing economic importance of pepper oleoresin is reflected in the worldwide production of dry pepper (*Capsicum*). The Food and Agriculture Organization (FAO) of the United Nations has reported that the production of dry peppers, which are the major raw material used in the production of oleoresin of *Capsicum*, increased by 35% from 2008 to 2018, with the production of 4 164 594 tonnes in 2018.[6]

The growing market for dry pepper is linked to the expanding production of *Capsicum* oleoresin. Oleoresin of *Capsicum* is primarily used by the food industry, mainly due to its sensory attributes of colour, flavour and taste.

In this industry, oleoresin of *Capsicum* is used as an ingredient in sauces, meat mixtures (*e.g.* sausages) and seasonings to confer a reddish colour or the famous pungency.[7] Although most people only know peppers for their sensory properties, the application of oleoresin of *Capsicum* in the food industry is not only for this purpose (Figure 8.2). It is also used as an additive to reduce oxidative processes in food, increasing its shelf life.

Pharmaceutical and cosmetic applications of oleoresin of *Capsicum* are also related to its inherent biological activity. Oleoresin of *Capsicum* is used as a nutritional and therapeutic additive. For instance, some muscular pain relief plasters and patches include oleoresin of *Capsicum* for its analgesic effect. Other products (*e.g.* shampoos and skin lotions) use oleoresin of *Capsicum* to deliver vitamins, antioxidants and antimicrobial compounds, which can improve hair growth, UV protection and skin health. An additional application is in weight control because capsaicinoids can reduce appetite and increase the use of calories by the body.[1,3,5] Oleoresin of *Capsicum* is used pharmacologically as a cardioprotector, antidiabetic agent, neuroprotector, anticancer agent and in the treatment of skeletal diseases (*e.g.* arthritis) (Figure 8.2).[8]

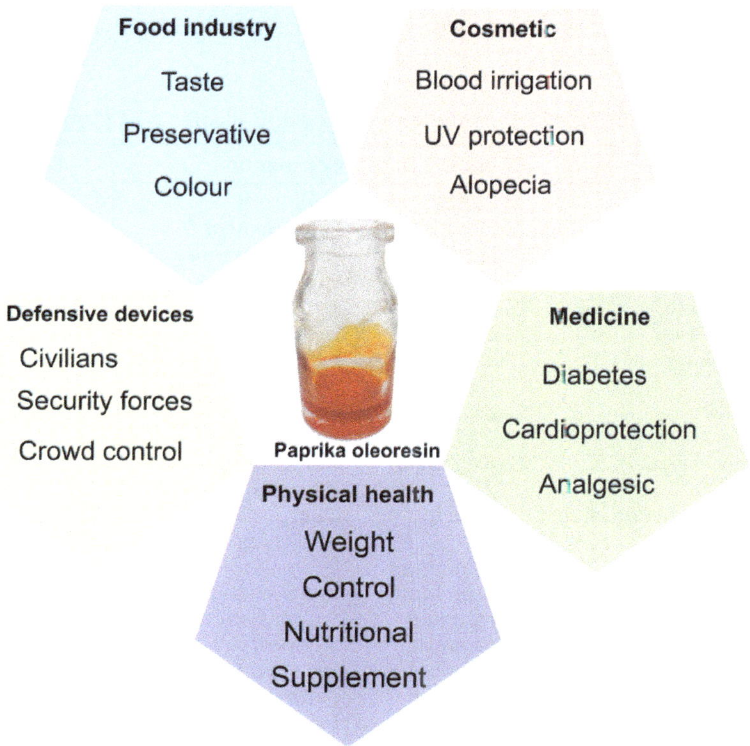

Figure 8.2 Commercial applications of *Capsicum* oleoresin. Image credit: Carlos Magno da Silva Antonio.

8.3 Characteristics of Oleoresins

Chapters 3 and 4 discussed how endogenous and exogenous factors affect the chemical composition of peppers. Similarly, the composition of the oleoresin is influenced by these factors. Each species, variety and cultivation method will produce a distinct oleoresin. Thus, depending on the desired chemical composition, oleoresin of *Capsicum* can be produced from a specific type of fruit or cultivation method. In general, oleoresin of *Capsicum* is produced in ripe fruits. At this stage of maturity, chilli peppers usually yield a higher amount of oleoresin than those in the green stage.[9]

The oleoresins are mainly composed of non-volatile substances, whereas the essential oils are primarily composed of volatile compounds. The oleoresin of *Capsicum* is rich in capsaicinoids, carotenoids, flavonoids, triglycerides, terpenes, vitamins A, C and E, fatty acids and minerals (Table 8.1)[10,11] and the concentration of such compounds in oleoresin of *Capsicum* will depend on their accumulation in the fruit. For instance, oleoresin of *Capsicum* produced from green cultivars or immature fruits will contain higher concentrations of phenolic compounds than capsaicinoids and carotenoids.[9,12]

As *Capsicum* peppers are mainly known by their diversity of colours and pungent taste, most of the oleoresins are prepared to have a high content of

Table 8.1 Major chemical compounds detected in oleoresin of *Capsicum*.

Compound	Chemical class	Compound	Chemical class
2-Decenal	Aldehyde	Oleic acid	Fatty acid
2-Heptanal	Aldehyde	Palmitic acid	Fatty acid
Decadienal	Aldehyde	Pentadecanoic acid	Fatty acid
Eicosane	Alkane	Steric acid	Fatty acid
Pentacosane	Alkane	Tricosanoic acid	Fatty acid
Pentadecane	Alkane	Farnesol	Fatty alcohol
Tetradecane	Alkane	Apigenin 6-*C*-pentoside-8-*C*-hexoside	Flavonoid
Capsaicin	Capsaicinoid	Luteolin 7-*O*-(2-apiosyl)-glucoside	Flavonoid
Dihydrocapsaicin	Capsaicinoid	Quercetin 3-*o*-glucoside-7-*o*-rhamnoside	Flavonoid
Homodihydrocapsaicin	Capsaicinoid	Quercetin 3-*O*-rhamonside	Flavonoid
Nonivamide	Capsaicinoid	β-Sitosterol	Sterol
Nordihydrocapsaicin	Capsaicinoid	Ergost-5-en-3-ol	Sterol
Capsanthin	Carotenoid	Stigmast-4-en-3-one	Sterol
Carotene	Carotenoid	Stigmast-5,22-dien-3-ol	Sterol
Cryptoxanthin	Carotenoid	Stigmast-5-en-3-ol	Sterol
Zeaxanthin	Carotenoid	β-Amyrin	Terpenoid
Phytol	Diterpene	Lupeol	Terpenoid
Arachidic acid	Fatty acid	Squalene	Terpenoid
Lignoceric acid	Fatty acid	Ascorbic acid	Vitamin
Linoleic acid	Fatty acid	Tocopherol	Vitamin

capsaicinoids and chromogenic compounds, particularly the red fraction carotenoids. Oleoresin of *Capsicum* will have a distinct composition depending on the cultivation process and extraction method. If the intended application is as a food additive, then the oleoresin of *Capsicum* cannot have a high content of capsaicinoids because they can promote toxic effects. According to the FAO, food-grade oleoresin of *Capsicum* must contain ≤0.5% of capsaicin and a residual solvent content <30 mg kg^{-1} of dichloromethane and acetone, 50 mg kg^{-1} of propanol, methanol and ethanol, and 25 mg kg^{-1} of hexane.[13] It is estimated that the oleoresin of *Capsicum* can be produced and enriched to present at least 18% of its composition as capsaicionoids.[14] Highly pungent oleoresins of *Capsicum* are not easy to commercialise because the excessive pungency is toxic on contact and therefore most commercial oleoresins of *Capsicum* are produced from *Capsicum annuum* and *Capsicum frutescens*, which are less spicy than *Capsicum chinense.*[15]

8.4 Extraction Methods

Several different methods are used to obtain oleoresins of *Capsicum*: conventional and non-conventional (Figure 8.3), solvent- and non-solvent-based. Conventional methods include only solvent-based techniques, created in ancient times and applied by elders to produce medicines. Conventional

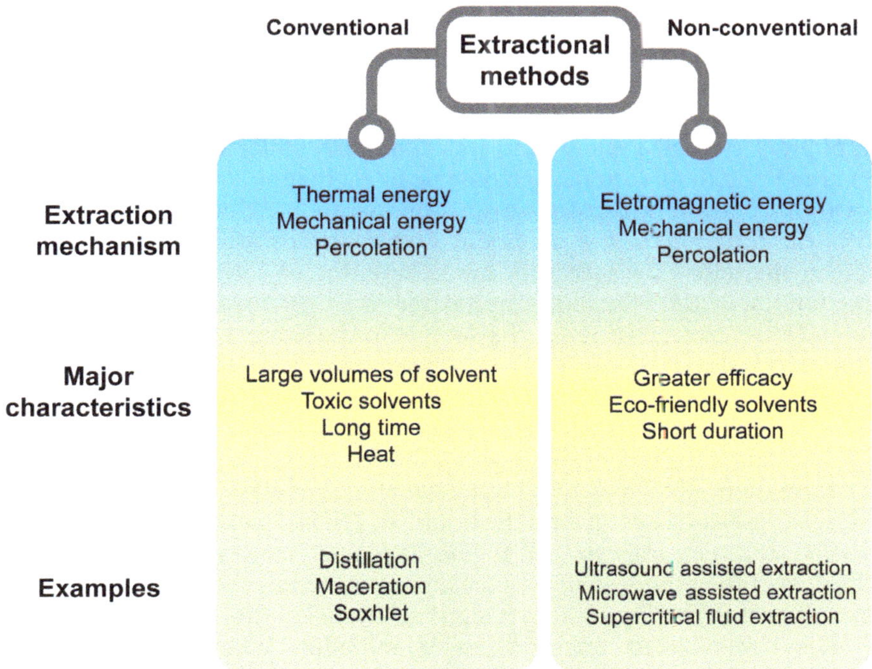

Figure 8.3 Features of conventional and non-conventional extraction methods. Image credit: Carlos Magno da Silva Antonio.

methods involve the extraction of chemical compounds from a sample based on the percolation of the solvent *via* solid–liquid or solid–vapour interactions (sample–solvent physical states). Such methods require long execution times, large amounts of solvent or high energy usage. By contrast, non-conventional methods are modern techniques that apply emergent technologies to make production more sustainable and eco-friendly. Non-conventional methods use technologies including microwave radiation, ultrasound or supercritical fluids to reduce the time, energy and solvents used.

Maceration is one of the most ancient of the conventional methods and is the same method as used to make tea or herbal infusions at home. The sample is left in contact with a solvent for a long period of time, with stirring of the sample or boiling of the solvent, or neither. After percolation or infusion, the extract is filtered, centrifuged or decanted to remove any solid residues. The most common solvents used are water, ethanol, methanol, hexane, acetone, acetonitrile, ethyl acetate, isopropanol or a mixture.[1,16,17] These solvents are also used in other extraction methods.

Regardless of the method applied, the solvent must be carefully chosen to optimise the extraction of the main chemical class required for the intended application of the oleoresin of *Capsicum*. For instance, if the desired metabolites are carotenoids, a non-polar solvent such as hexane is recommended; for flavonoids, which are more polar and hydrophilic, solvents such as water and hydroalcoholic mixtures are suitable; and, if capsaicinoids are the goal, polar aprotic solvents such as acetone, acetonitrile and ethanol can be usen. Solvents with low toxicity, such as water and ethanol, are preferable for applications involving ingestion or direct contact with the eyes, mouth or respiratory system. Ethanol and water are also easier to handle on an industrial scale than acetone, acetonitrile or methanol.[1,16,17]

Other oleoresin of *Capsicum* extraction methods include those assisted by ultrasound or microwave radiation. Both methods are also referred to as "green methods" because they lower amounts of solvent and shorter extraction times. Ultrasound-assisted extraction (UAE) uses ultrasonic waves to enhance the mass transfer of the sample to the solvent. The sample can be exposed to ultrasonic waves *via* the insertion of a probe or using an ultrasonic bath. UAE operates *via* cavitation. The propagation of an ultrasonic wave through a liquid causes a series of modifications in the pressure of the liquid by inducing compression and rarefaction in its molecules. This causes the formation, expansion and implosion of microbubbles. The bubbles then collapse, releasing energy into the liquid, which induces particle breakdown, erosion, the disruption of cell membranes and micro-stirring (Figure 8.4). The area of the liquid–solid contact surface is increased and cell metabolites are exposed to the liquid, favouring their diffusion into the solvent.[18,19]

Microwave-assisted extraction uses radiation with frequencies from 2.54 GHz to 915 MHz. This microwave energy can directly affect the electron cloud of a molecule, but not its molecular structure. Microwaves promote dipole polarisation and ionic conduction. Dipole polarisation occurs

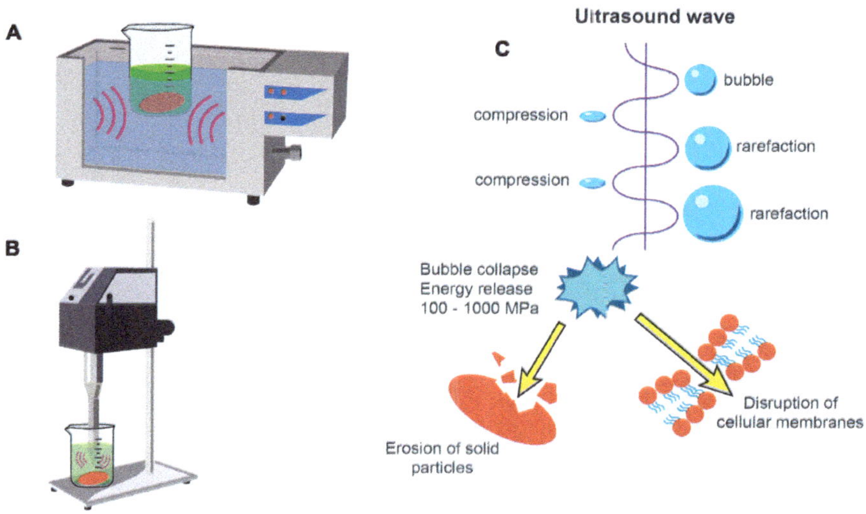

Figure 8.4 Schematic diagrams of (A, B) the instruments used in ultrasound-assisted extraction and (C) the mechanism of action. Image credit: Carlos Magno da Silva Antonio.

because the electrical field of the microwaves aligns with the molecule's dipole and cause it to rotate. When the electrical field is removed, the molecule returns to its original dipole alignment and the energy is released as heat. Ionic conduction refers to the migration of ions under the effect of the electromagnetic field of the microwaves. The movement of ions in an alternating electromagnetic field generates frictional heat with the media and by collision with other migrating ions (Figure 8.5). The heat processes in microwave-assisted extraction initiate uniformly inside the matrix, avoiding the thermal degradation observed in conventional methods. The energy produced vaporises the liquid inside cells, increasing the intracellular media and disrupting the pores of the cellular membrane, enhancing mass transfer from the intracellular to extracellular media.[1,20,21]

Supercritical fluid extraction (SFE) and cold compression extraction are used to reduce solvent residues within the oleoresin of *Capsicum*. Both methods require low-temperature heating of the sample or solvent, preserving thermosensitive compounds. In SFE, the extraction solvent is a supercritical fluid, which means that, during extraction, the pressure and temperature are modified above critical points such that the boundaries between the liquid and gas phases disappear. This can increase the permeability of the solvent and its selectivity in solvating other analytes. SFE usually uses solvents that are gases under normal temperatures and pressures, facilitating the removal of the solvent after the extraction procedure and improving yields. Most SFE methods use carbon dioxide as the solvent because its supercritical point is at 32 °C and 7.4 MPa.[22,23]

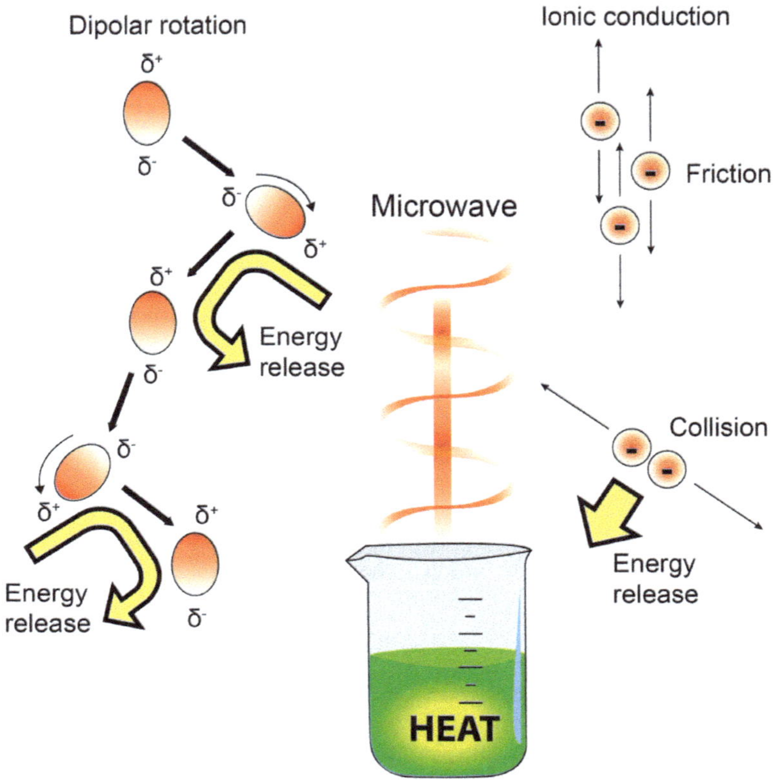

Figure 8.5 Schematic diagram showing the generation of heat in microwave-assisted extraction. Image credit: Carlos Magno da Silva Antonio.

Cold compression or cold pressing does not involve the use of a solvent; the sample is simply mechanically pressed to force out any liquid. This method can be used to produce oleoresin of the seeds from agricultural or industrial residues of pepper cultivation. Cold compression is more applicable to the direct infusion of pepper metabolites into another essential oil (*e.g.* extra virgin olive oil), but is usually associated with low extraction yields.[1,24]

8.5 Comparison of Extraction Methods for the Production of Oleoresin of *Capsicum*

Before you think, "Finally, I will know which is the best extraction method!", you must be aware of some common ideas in phytochemistry and agro-industrial development. The selection of the extracting solvent depends on the aim of the final product and the same applies to the choice of method. Several other factors will also affect this decision, such as the implementation cost, the demand for the product (small, medium or large industrial), the objective of the extraction (scientific research, industrial production or the isolation of high-purity fractions), the desired characteristics of the final product, and the target market and its regulations.

For instance, paprika oleoresin can be obtained by any of the described methods if it meets the physical and chemical standards defined by the FAO for classification as a food-grade product. Different chemical profiles will be obtained depending on the extraction method (Table 8.2). Highly nutritional oleoresins of *Capsicum* can be obtained using SFE, although this is only economically profitable for medium- to large-demand industries. UAE can yield similar results to SFE, but with the advantages of requiring cheaper equipment and being easier to implement. However, neither of these two methods is used by small or local producers because maceration and cold compression involve less exposure of employees to solvents and require less technical skill and training to operate the machinery.[11,24,25]

If the desired chemical profile of the oleoresins of *Capsicum* includes volatile compounds (generally highly antioxidant compounds), then it is preferable to use extraction methods that are not dependent on high-temperature applications, such as Soxhlet extraction. UAE is one of the most commonly used methods in *Capsicum* exploitation due to the versatility of the solvents, the low temperatures required, low cost, rapid execution, high recovery rates and repeatability.

As oleoresin of *Capsicum* can have several different purposes, let us give some examples. *C. chinense* varieties have extremely high levels of capsaicins in their oleoresin (Table 8.2) and are therefore useful in products requiring pungency, such as deterrent devices intended for self-defence/riot control, or for the isolation of capsaicins intended for medicines. *C. annuum* cultivars are less pungent and thus more suitable for use as food additives, nutritional products and spices.

8.6 Concluding Remarks

The oleoresin of *Capsicum* is a remarkable bioproduct that concentrates several of the biological and sensory properties of peppers. It is a versatile product used in various industrial sectors. Its major application is as a food additive, mainly due to the antioxidative properties of its carotenoids and capsaicinoids, its colour and its pungency, which preserve products and remove insipidity. Other important chemical characteristics include the presence of vitamins, flavonoids, fatty acids and reducing sugars. These chemical classes contribute to antioxidative and antidiabetic properties of oleoresin of *Capsicum*.

Despite the numerous applications of oleoresin of *Capsicum*, its production must be designed specifically for its intended final use. The chemical composition of the oleoresin, and consequently its pharmaceutical properties, are influenced by the species, cultivar, maturity stage, cultivation methods, environmental conditions of growth and extraction method. The production of oleoresin of *Capsicum* can be achieved by conventional or

Table 8.2 Summary of methods used to extract oleoresin of *Capsicum*.

Method	Extraction time	Temperature (°C)[a]	Remarks	Species	Cultivar/variety	Extraction solvent	Capsaicin (mg kg^{-1})	Total phenolics	Carotenoids	Flavonoids	Ref.
Conventional methods											
Soxhlet extraction	4–6 h	90	Uses boiling solvent, long extraction times, energy consumption 21 600 kJ (300 min), thermal decomposition, low selectivity	*C. baccatum*	Dedo de moça	Hexane	1440	0.82[c]	—	—	26
				C. frutescens	Malagueta	Hexane	42	36[c]	—	—	22
Maceration	30 min–5 days	25–65	High consumption of solvent, long extraction time, low selectivity	*C. annuum*	Jinda	95% ethanol	7000	—	—	—	24
				C. annuum	Superhot	95% ethanol	11 900	—	—	—	24
				C. chinense	Habanero orange	80% Methanol	110 400	—	—	—	3
				C. chinense	—	Ethanol	—	277.62[d]	—	—	27
Non-conventional methods											
Ultrasound-assisted extraction	10 min–2 h	4–65	Energy consumption 102 kJ (20 min), low execution time, versatile application of solvent, inexpensive and simple, higher efficacy than traditional methods	*C. baccatum*	Dedo-de-moça	Methanol	120	6.33[c]	—	—	26
				C. baccatum	Dedo-de-moça	Ethanol	130	7.89[c]	—	—	26
				Commercial sample	—	—	—	—	91 806[e]	—	28
				C. chinense	Thapthim Mo Diadaeng	80% Methanol	13.62	7.15[b]	273.76[b]	14.33[b]	3
				C. chinense	Phirot	80% Methanol	25.97	6.392[b]	350.12[b]	23.110[b]	3
				C. frutescens	Malagueta	Methanol	984	1.7[c]	—	—	29
				C. chinense	Naga jolokia	Methanol	4457	3.1[c]	—	—	29
Microwave-assisted extraction	1–5 min	65–125	Energy consumption 384 kJ (20 min), low execution time, versatile application of solvent, higher efficacy than traditional methods, lower recovery of non-polar or volatile compounds	*C. annuum*	Cayenne	Ethanol	6730	—	—	—	30

Supercritical fluid extraction	90 min	40–60	Uses low-toxicity extraction solvents, low solvent residues in the final product, highest extraction recovery among the methods discussed here, absence of time-consuming process for sample concentration/solvent elimination	*C. frutescens*	Malagueta	CO_2	115	—	—	—	11
				C. frutescens	Malagueta	CO_2	30	30[c]	—	—	22
				C. annuum	Jalapeño	CO_2	12.18	6.39[b]	350.1[b]	23.11[b]	31
				C. annuum	Jalapeño	Methanol	3.4	0.6[c]	—	—	29
				C. baccatum	Dedo de moça	Methanol	280	0.8[c]	—	—	29
Cold compression	—	25–40	No extraction solvent, low cost of implementation, low extraction temperature, unpopular with customers, few published descriptions for *Capsicum*, electricity consumption 1 kWh (1 h), applicable in small- and medium-sized industries	*C. annuum*	Jinda	—	22 300	—	—	—	24
				C. annuum	Superhot	—	26 400	—	—	—	24

[a]Temperature (°C) = extraction temperature; Ref. = references.
[b]mg CE kg^{-1} dry weight.
[c]mg GAE g^{-1}.
[d]mg GAE 100 g^{-1}.
[e]μg g^{-1}.

unconventional methods. Each method has its own advantages and disadvantages, which must be addressed based on the production objectives. For instance, maceration is easily implemented in large-scale production, although they incorporate more solvent residues into the bioproducts than unconventional methods, such as SFE. The design of the experiment is crucial for this versatile and highly chemically complex natural product.

References

1. G. Melgar-Lalanne, A. J. Hernández-Álvarez, M. Jiménez-Fernández and E. Azuara, *Food Bioprocess Technol.*, 2017, **10**, 51–76.
2. A.-J. Hernández-Álvarez, M. Mondor and G. Melgar-Lalanne, in *Cold Press Oils*, ed. M. F. Ramadan, Academic Press, London, 1st edn, 2020, **58**, pp. 665–675.
3. P. Sricharoen, N. Lamaiphan, P. Patthawaro, N. Limchoowong, S. Techawongstien and S. Chanthai, *Ultrason. Sonochem.*, 2017, **38**, 629–639.
4. L. Koleva Gudeva, V. Maksimova, M. Serafimovska Darkovska, R. Gulaboski and E. Ivanovska Janevik, *Sci. Work. Food Sci. Eng. Technol.*, 2013, **60**, 917–922.
5. N. Baenas, M. Belović, N. Ilic, D. A. Moreno and C. García-Viguera, *Food Chem.*, 2019, **274**, 872–885.
6. Food and Agriculture Organization of the United Nations, http://www.fao.org/faostat/en/#data/QC, acessed February 2021.
7. J. Echave, A. G. Pereira, M. Carpena, M. Á. Prieto and J. Simal-Gandara, in *Capsicum*, ed. A. Dekebo, IntechOpen, London, 2nd edn, 2020, DOI: 10.5772/intechopen.91785.
8. M. Lu, C.-T. Ho and Q. Huang, *J. Food Drug Anal.*, 2017, **25**, 27–36.
9. F. Conforti, G. A. Statti and F. Menichini, *Food Chem.*, 2007, **102**, 1096–1104.
10. J. P. Fernández-Trujillo, *Grasas Aceites*, 2007, **58**, 252–263.
11. A. C. de Aguiar, J. F. Osorio-Tobón, L. P. S. Silva, G. F. Barbero and J. Martínez, *J. Supercrit. Fluids*, 2018, **133**, 86–93.
12. A. Wesołowska, D. Jadczak and M. Grzeszczuk, *Acta. Sci. Pol. Hortorum Cultus*, 2011, **10**, 171–184.
13. Food and Agriculture Organization of the United Nations, Prepared at the 35th Expert Committee on Food Additives Meeting (1989), published in Compendium of Food Additive Specifications 49 (1990).
14. K. G. Liljana, M. Viktorija, S. D. Marija, R. Gulabovski and I. J. Emilija, *Food Sci. Eng. Technol.*, 2013, **LX**, 917–922.
15. N. Riquelme and S. Matiacevich, *CYTA-J. Food*, 2017, **15**, 344–351.
16. R.-M. Stoica, M. Moscovici, C. Tomulescu and N. Babeanu, *Sci. Bull. Ser. F. Biotechnol.*, 2016, **20**, 93–98.
17. A. S. Antonio, L. S. M. Wiedemann and V. F. da Veiga Junior, *Food Anal. Methods*, 2019, **12**, 1327–1345.
18. F. Chemat, N. Rombaut, A. G. Sicaire, A. Meullemiestre, A. S. Fabiano-Tixier and M. Abert-Vian, *Ultrason. Sonochem.*, 2017, **34**, 540–560.

19. B. K. Tiwari, *Compr. Anal. Chem.*, 2015, **76**, 111–129.
20. T. Belwal, F. Chemat, P. R. Venskutonis, G. Cravotto, D. K. Jaiswal, I. D. Bhatt, H. P. Devkota and Z. Luo, *TrAC, Trends Anal. Chem.*, 2020, **127**, 115895.
21. N. Flórez, E. Conde and H. Domínguez, *J. Chem. Technol. Biotechnol.*, 2015, **90**, 590–607.
22. A. C. De Aguiar, L. P. Sales, J. P. Coutinho, G. F. Barbero, H. T. Godoy and J. Martínez, *J. Supercrit. Fluids*, 2013, **81**, 210–216.
23. M. Yousefi, M. Rahimi-Nasrabadi, S. M. Pourmortazavi, M. Wysokowski, T. Jesionowski, H. Ehrlich and S. Mirsadeghi, *TrAC, Trends Anal. Chem.*, 2019, **118**, 182–193.
24. L. Charoenchai, T. Chusut, T. Wunnakup and A. Sukhonthapun, presented in part at RSU International Research Conference 2019, 2019, pp. 268–277.
25. M. F. Fabela-Morón, J. C. Cuevas-Bernardino, T. Ayora-Talavera and N. Pacheco, *Food Rev. Int.*, 2020, **36**, 105–134.
26. A. L. B. Dias, C. S. Arroio Sergio, P. Santos, G. F. Barbero, C. A. Rezende and J. Martínez, *J. Food Eng.*, 2017, **198**, 36–44.
27. L. S. Santos, C. C. Fernandes, L. S. Santos, I. P. B. Deus, T. L. De Sousa, M. Lazaro and D. Miranda, *Food Sci. Technol.*, 2020, **2061**, 1–8.
28. D. Hornero-Méndez and M. I. Minguez-Mosquera, *J. Agric. Food Chem.*, 2001, **49**, 3584–3588.
29. A. C. De Aguiar, J. P. Coutinho, G. F. Barbero, H. T. Godoy and J. Martínez, *Int. J. Food Prop.*, 2016, **19**, 1292–1302.
30. F. Nazari, S. N. Ebrahimi, M. Talebi, A. Rassouli and H. R. Bijanzadeh, *Phytochem. Anal.*, 2007, **18**, 333–340.
31. P. Grande-Villanueva, A. C. De Aguiar, J. Pereira-Coutinho, H. Teixeira-Godoy, E. M. Escamilla-Silva and J. Martinez, *Cienc. Tec. Vitivinic.*, 2015, **30**, 79–104.

CHAPTER 9

Analytical Methods Applied to Capsicum

9.1 Introduction

Previous chapters have shown how the specialised metabolite content of *Capsicum* peppers varies and the importance of understanding these modifications to correctly explore the biodiversity of this genus. But how can we monitor the chemical profile of *Capsicum*? What is the best way to extract the metabolites? These are only a few questions that arise when researching this peculiar genus.

The objective of the research task must be clear before selecting and comparing analytical methods. Several methods are available for the chemical characterisation of *Capsicum* and their selectivity, sensitivity and accuracy differ. Some methods will produce a complete chemical profile, whereas others only detect a specific compound or chemical class. In some cases, concentrations as low as parts per billion can be quantified, whereas other methods only give qualitative information. It important to remember that no specific characteristics make a single method the one of choice over all others for every situation: this depends on the objective of the research. It is not always necessary to use ultra-high-performance methods, just as it is not necessary to use a scalpel to slice bread. Be sure to choose the method more suitable for your goal and remember that the experimental design is fundamental. The methods used to analyse the chemical composition of chilli peppers range from organoleptic methods to high-performance techniques, such as high-resolution mass spectrometry and nuclear magnetic resonance (NMR) spectrometry.

Food Chemistry, Function and Analysis No. 37
Chemistry and Nutritional Effects of Capsicum
By Valdir Florencio da Veiga, Jr, Larissa Silveira Moreira Wiedemann,
Claudio Pereira de Araujo, Jr and Ananda da Silva Antonio

Published by the Royal Society of Chemistry, www.rsc.org

Chapter 8 discussed the major extraction methods used in the preparation of *Capsicum* samples, so we strongly recommend you read that chapter first. This chapter takes a closer look at the methods most commonly used to analyse *Capsicum*.

9.2 Scoville Heat Test

The Scoville heat test was the first effective analytical method developed to classify chilli peppers by their pungency. Created in 1912 by Wilbur Scoville, it is based on oral experimentation. The Scoville heat test expresses the pungency as Scoville heat units (SHU), a term that you have probably heard before. In this test, the chilli pepper sample is ground and extracted overnight with 95% ethanol at a proportion of 1 : 100 (g : mL). The resulting extract is serially diluted with a 5% sucrose solution. The dilutions are then given to a panel of five trained tasters, who must immediately swallow the sample and record whether or not it is pungent. There must be a time lapse of 30 min between testing each dilution. The test is considered to be over when at least three of the tasters register no pungency in the dilution. The SHU value is therefore the dilution factor required to extinguish the pungency of the chilli pepper extract. SHU are measured as multiples of 100 units[1–3] and currently vary from 0 to about 3 000 000 (for Carolina Reaper, the hottest pepper in the world since 2019).

The heat scale based on SHU is still used today, although the test itself became obsolete about 10 years after its creation. The Scoville method was extensively criticised by the scientific community as a result of several factors: the lack of a reference standard; a high influence from human error because it depends on an individual's sensitivity to pungency; a lack of reproducibility; a build-up of the heat sensation in prolonged tests; and a rapid loss of tasters due to tongue desensitisation. However, as it was, and still is, a famous method, several researchers developed instruments capable of determining the SHU, but without the problems inherent in the organoleptic test.

9.3 Spectrophotometric Methods

Spectrophotometric methods analyse samples through their interaction with electromagnetic energy, including radiation in the visible, ultraviolet (UV) and infrared ranges. These methods are usually performed in the presence of a chromogenic reactant, in which case they are also known as colorimetric methods. Several spectrophotometric methods can be applied to *Capsicum* samples, mainly to detect and quantify the capsaicinoids, total phenolic compounds, flavonoids and antioxidants.[4–7]

The application of colorimetric methods must be performed carefully. First, the analyst must remember that these methods only give a total content and are unable to distinguish between the analyte and interfering compounds, or even give specific information about each compound present in the sample. Spectrophotometric methods give information regarding a

Table 9.1 Chromogenic reactions used in the analysis of *Capsicum* fruits.

Chromogenic reagent	Reaction type
2,6-Dichloroquinone-4-chloroimide	Condensation with phenols
Folin–Ciocalteu assay	Reduction/oxidation reaction with phenols in alkaline solution
Vanadium chloride	Nitrosation of the vanillyl group
Sodium nitrate/sodium molybdate	Nitration of the vanillyl group
Ferric chloride	Reduction/oxidation reactions with antioxidants
Phosphomolybdic acid	Reduction/oxidation reactions with phenols
4-(2-Pyridylazo) resorcinol	Complexation reaction

particular chemical class (Table 9.1). Their ongoing application, even with the availability of high-performance methods, is due to their low cost, simplicity and rapid execution. If the goal is to determine the total content of a specific chemical class quickly, then these methods are suitable. The application of colorimetric methods is mainly related to quality monitoring of production lines, preliminary evaluations and prospective studies.[5,8–10]

The disadvantages of colorimetric methods include the high toxicity of chromogenic reagents and the discard volume produced over long time periods. For these reasons, most colorimetric methods are not considered to be green or sustainable chemistry techniques. There are also technical issues that need to be addressed. Colorimetric methods currently only provide information about a particular chemical class. This is a question of selectivity, which can only be solved with sample pre-treatment to purify or separate the target analyte.[11] In colorimetric methods, sample pre-treatment can be carried out during extraction with liquid–liquid partitioning or by chromatographic methods, such as thin-layer chromatography (TLC). In TLC, the chromogenic reactant can be used as a dye reagent for the stains in the TLC plates. The application of TLC also increases the selectivity by using standard solutions and comparing the retention factor of each stain on the plate.[6,11–13] Combinations of separation and colorimetric methods can quantify analytes through absorbance and TLC-densitometry techniques.[11,13–16]

Another technical feature that needs to be addressed is the working range. Ultraviolet and visible quantification methods are based on the Beer–Lambert law, but are only valid within a certain concentration range. Highly concentrated samples will not obey the Beer–Lambert law because intermolecular interactions will interfere in the absorbance of the sample. The linearity range varies depending on the composition of the sample and the instrument and can only be determined experimentally. The linearity range in the colorimetric analysis of capsaicinoids varies from 50 to 1000 $\mu g\,mL^{-1}$.[5,6,11]

9.3.1 Ultraviolet and Near-infrared Methods

The major chemical classes present in *Capsicum* can be analysed using UV and near-infrared (NIR) electromagnetic radiation. Carotenoids, flavonoids

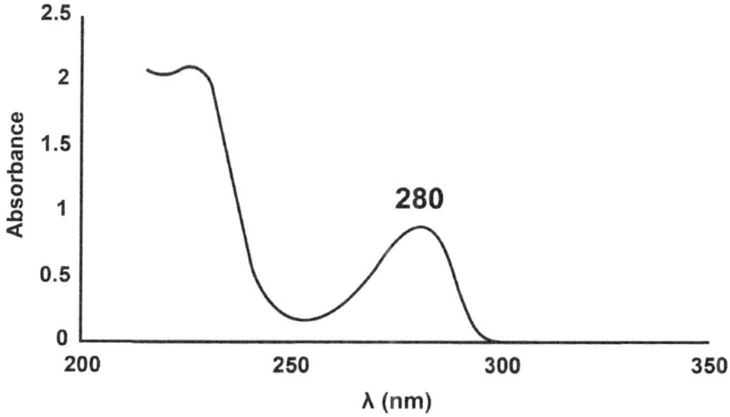

Figure 9.1 UV absorbance spectrum of capsaicinoids. Image credit: Carlos Magno da Silva Antonio.

and capsaicinoids can be quantified without being destroyed, as occurs in colorimetric methods. Because capsaicinoids are not found in any genera other than *Capsicum*, we focus on their analysis for the rest of this chapter. UV methods for the determination of capsaicinoids are performed at 280 nm for absorbance and 320 nm for fluorescence (Figure 9.1).[17] Several natural metabolites, such as flavonoids, carotenoids and other classes of alkaloids, can also interact with UV radiation at the same wavelength as capsaicinoids. UV detection is therefore mainly carried out after separation by chromatographic methods, such as high-performance liquid chromatography (HPLC) and TLC.[11]

Depending on the efficacy of the chromatographic separation (the chromatographic resolution), the UV detector can give accurate results with lower detection and quantification limits than colorimetric methods (Table 9.2). This detection method is therefore more sensitive for the detection of capsaicinoids, although it requires a more selective extraction of the sample or separation of its constituents.

Near-infrared methods are a reliable option to detect capsaicinoids mainly because of their low cost and the fact that previous extraction is not required. NIR methods can be used to quantify capsaicinoids within solid samples, including pepper powder or even the whole fruit.[24,25] This detection method was the first method reported to be capable of distinguishing capsaicin from its natural analogue, nonivamide, by the presence of strong bands at 970 and 1040 cm^{-1}.[26]

The determination of capsaicinoids by NIR is usually performed within the band ranges 3600–2800 and 1800–600 cm^{-1}, which are related to the functional groups –OH, –C–O and –C=O. Flavonoids, carotenoids and water can be interferents. This method therefore requires dry samples and chemometric analysis to process the significant amount of data produced by Fourier transform NIR spectrometry (FT-NIRS).[25,27,28] Despite these

Table 9.2 Characteristics of UV detection methods applied to the determination of capsaicin.[a]

Detection method	Sample preparation	λ (nm)	Linear range	Sample	LOD	Y (%)	Ref.
Absorbance	Solid–liquid extraction; HPLC	280	10–100 mg mL^{-1}	Dried fruit	0.161 mg mL^{-1}	98.5–99.3	18
Absorbance	UAE; HPLC	280	0.0693–346.5 ppm	Frozen fruit	0.1242 ppm	>95	19
Absorbance	MAE; UHPLC	200–400	—	Frozen fruit	0.556 mg L^{-1}	—	20
Absorbance	UAE; Solid-phase extraction; HPLC	200–400	1–1000 mg L^{-1}	Dried fruit	1.65 mg L^{-1}	98.91	21
Absorbance	Maceration; HPLC	230	1–80 mg mL^{-1}	Dried fruit	148.34 ng mL^{-1}	95–98	22
Absorbance	Maceration; UHPLC	280	0.625–500 mg mL^{-1}	Dried fruit	0.040 mg mL^{-1}	94	23
Fluorescence	Maceration; UHPLC	280; 310	0.3–75 mg mL^{-1}	Dried fruit	0.136 mg mL^{-1}	95–98	23
Fluorescence	Solid-phase extraction; HPLC	320	1.0–50 ng mL^{-1}	Paprika	0.29 ng mL^{-1}	93–107	17

[a]LOD = limit of detection; Ref. = reference; Y (%) = extraction yield (%); HPLC = high-performance liquid chromatography; UAE = ultrasound-assisted extraction; MAE = microwave assisted extraction; UHPLC = ultra-HPLC.

interferents, NIR spectrometry can quantify capsaicin in pepper samples in powder form with the same sensitivity as an HPLC-UV method, which requires expensive separation steps.[25,29,30] The major application of NIR is currently in the characterisation of isolated capsaicinoids, quality control in industry (as a non-destructive method) for production lines and the surveillance of adulterated products.

9.4 Chromatographic Methods

Chromatography is widely used in the research and development of natural products. It is a set of physicochemical separation methods based on differential migration. A chromatographic system consists of two immiscible phases in intrinsic contact, through which the sample will pass. These immiscible phases are known as the stationary phase and the mobile phase. The stationary phase is deposited on a fixed support, while the mobile phase passes through it carrying the sample. Separation occurs as each compound in the sample interacts differently with these two phases, requiring a different time to cross the chromatographic system: hence the term differential migration (Figure 9.2). As this process proceeds, the compounds in the sample are separated, facilitating their individual analysis.

Chromatographic methods alone do not provide any information about the chemical composition of the sample or the molecular structure of the separated compounds. The only information that chromatography provides is the retention time, the elution time or the retention factor, which are measures of how long each compound takes to cross the chromatographic system. If a standard solution is used under the same chromatographic conditions and in the same system, then its retention time or factor can be used to estimate its presence in a real sample. To provide specific information on the composition of a sample, a detection method must be used alongside chromatography. Chromatographic methods are extremely

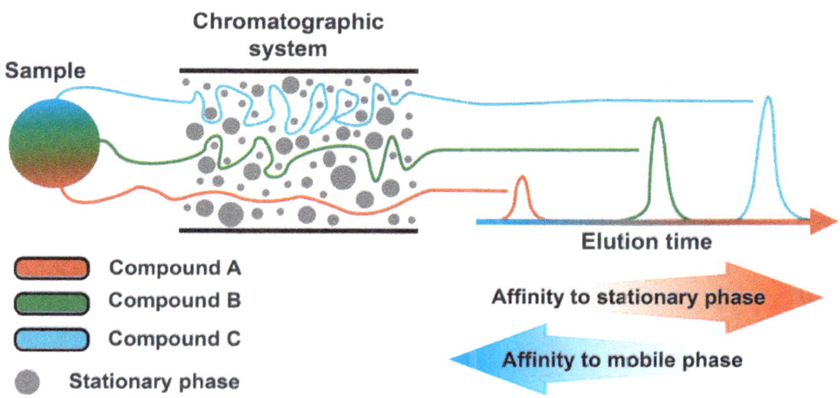

Figure 9.2 Differential migration processes in chromatography. Image credit: Carlos Magno da Silva Antonio.

versatile and can be used in conjunction with other techniques that provide such data. For instance, UV methods are reliable detectors for capsaicinoids if used with an appropriate separation method.

The main concern here is in choosing the best chromatographic system to separate metabolites within the sample. Several combinations of the mobile and stationary phases can be used in different types of chromatographic system. TLC is mostly used with colorimetric methods and is low cost, reproducible, rapid and easy to perform. When a UV method is used instead of colorimetric detection, TLC can be used to isolate the pepper metabolites. Chromatography can be applied to a large variety of extracts, ranging from fresh powdered fruits to synthetic drugs. The working range of TLC is comparable with those of UV methods (Table 9.2) and can determine compounds from 50 to 1000 mg mL^{-1}.

Capsicum metabolites, such as capsaicinoids, flavonoids and carotenoids, are generally analysed with normal-phase TLC. This means that the stationary phase of the TLC system is polar, usually a silica gel coated with the fluorescent indicator F254 to enable visualisation of the spot by UV detection.[11] The mobile phase will have a low polarity (Table 9.3). Low-polarity mobile phases increase the efficacy of TLC separation in *Capsicum* samples. A mobile phase of toluene/acetone/chloroform (45 : 30 : 25 v/v/v) often gives the best resolution in the separation of capsaicinoids. The resolution will depend on the characteristics of the sample, including the sample composition. Another mobile phase may be more suitable for different types of matrix. It is important to evaluate the best option on a case-by-case basis.[6]

As the chemical complexity of the samples increases, TLC cannot be used to provide specific data, such as distinguishing each compound present or even separating capsaicinoids (a type of alkaloid) from other chemical classes. If qualitative and quantitative data are required for each capsaicinoid present, comprehensive chemical profiles or even the isolation of a metabolite, then high-performance chromatographic methods must be used, such as HPLC or gas chromatography.

The most commonly used chromatographic method in *Capsicum* analysis is HPLC due to the high versatility of the chromatographic systems that can be applied. Several mobile and stationary phases can be combined to provide the best resolution (Table 9.4). In addition, spectrometric and spectrophotometric methods can be used for detection. HPLC usually uses C18 columns due to their low polarity and selectivity to compounds with acyl chains. With this reversed phase, the major capsaicinoids are more strongly retained in the chromatographic system with greater retention times. For the minor capsaicinoids, C30 columns or phenyl stationary phases are commonly used to increase selectivity.

Despite the incredible selectivity and sensitivity of HPLC, it are not a popular option for large-scale experiments. This is because each analysis is time consuming, varying from 30 to 40 min. To overcome this, ultra-HPLC (UHPLC) can be applied. UHPLC uses smaller columns with smaller particle sizes, lower flow rates and higher pressures to enable chromatographic runs

Table 9.3 Thin-layer chromatography mobile phases used in the analysis of *Capsicum*.

Mobile phase	Proportion (v/v)	Retention factor		Ref.
Ether : acetate : methanol	75 : 20 : 5	0.98	β-Carotene	31
		0.31	CAPS[a]	
Hexane : acetone	4 : 1	0.44	Stigmasterol	32 and 33
Cyclohexane : chloroform : acetic acid	70 : 20 : 10	0.16	CAPS	16
		0.38	Vanillin	
		0.22	Ferulic acid	
		0.10	Coumaric acid	
Toluene : ethyl acetate	6 : 4	0.38	CAPS	34
Hexane : ethyl acetate	50 : 50	0.41	CAPS	35
	1 : 1	0.30	CAPS	
		0.85	Carotenoids	
Ethyl ether : chloroform : acetonitrile	40 : 45 : 15	0.68	CAPS	6
Toluene : acetone : chloroform	45 : 30 : 25	0.56	CAPS	
Chloroform : methanol : acetic acid : hexane	2.85 : 0.15 : 0.15 : 1	0.79	Capsaicin	36
Petroleum ether : chloroform : acetonitrile	40 : 45 : 15	0.078	Capsaicin	8
Chloroform : methanol : acetic acid	9.5 : 0.5 : 0.1	0.52	Capsaicin	14
Petroleum ether : ethyl acetate	90 : 10	0.13	Terpenoids	37
		0.20	Alkaloids and phenols	
		0.33	Alkaloids and terpenoids	
		0.40	Alkaloids and terpenoids	
		0.47	Alkaloids	
		0.60	Phenols and terpenoids	
		0.67	Anthraquinone	
		0.80	Alkaloids and terpenoids	
Hexane : ethyl acetate	6 : 4	0.32	Capsaicin	38

[a]CAPS = capsaicinoids.

as fast as 6 min. Examples of other stationary phases that can be used in HPLC include ethylene-bridged hybrid and fused-core columns.[19,20] Within HPLC and UHPLC, the working range varies from 0.0055 to 365 mg mL^{-1}.[11] Although the maximum limit of this working range is lower than other techniques, the sensitivity is increased, enabling the determination of minor constituents. The limit of detection and quantification in HPLC can be as low as 0.001 and 0.003 μg mL^{-1}, respectively.

The major application of HPLC in *Capsicum* analysis is due to the great interest in capsaicinoids and carotenoids. However, if the intended purpose is to determine volatile organic compounds, then gas chromatography is more suitable. Gas chromatography uses an inert gas, which does not interact with the sample, as the mobile phase. Separation occurs *via* the interaction of the sample with the stationary phase and the progressive volatilisation of the sample metabolites during a thermal programme. The

Table 9.4 Chromatographic conditions applied in the analysis of *Capsicum*.[a]

Sample	Sample preparation	Method	Stationary phase	Mobile phase	Flow rate	Detection	Ref.
Pericarp of fresh fruit	Maceration with acetone : petroleum ether (1 : 1)	HPLC	C18	Linear gradient from 0 to 100% of solvent A (10% acetonitrile in water) and solvent B (90% acetonitrile)	1 mL min^{-1}	UV	31
Powdered fruit	UAE with acetonitrile followed by solid-phase extraction (C18)	HPLC	RP-8	Gradient of methanol in water ranging from 55 to 70% of methanol	0.7 to 1.2 mL min^{-1}	UV	21
Dried fruit	Maceration with ethyl acetate	HPLC	ODS Hypersil	Isocratic elution of 70% aqueous methanol	1 mL min^{-1}	UV	15
Dried fruit	Maceration with acetonitrile	HPLC	C18 (4.6×250 m; 5 cm^3)	Gradient elution of methanol in water from 70 to 100% methanol	2 μL min^{-1}	MS	39
Fresh fruit	Maceration with methanol, followed by acid–base partitioning	GC	SE30 (30 m×0.25 mm)	Gradient temperature from 60 °C to 280 °C (16 °C min^{-1})	1 mL min^{-1}	MS	40
Analgesic (topical formulation)	Liquid–liquid extraction with methanol : water (60 : 40 v/v)	LC	C18 (250×4.6 mm; 5 μm)	Gradient elution of A [methanol : water : phosphoric acid 1M (47 : 43 : 10 v/v/v)] and B [methanol : water : phosphoric acid 1M (67 : 23 : 10 v/v/v)]	1 mL min^{-1}	UV MS	41
Fresh fruit	Maceration with acetone followed by partitioning with salting-out effect	HPLC	C30 (250×4.6 mm; 5 μm)	Gradient elution of A [methanol : MTBE : water (82 : 16 : 2 v/v/v)] and B [methanol : MTBE : water (10 : 88 : 2 v/v/v)]	0.8 mL min^{-1}	UV APCI MS	42
Fresh fruit	UAE with methanol : acetone	UHPLC	Hypersil Gold C18 (3×100 mm; 1.9 μm)	Isocratic elution with water : acetonitrile 40 : 60 v/v	0.5 mL min^1	APCI MS	43
Fresh fruit	UAE with acetonitrile	GC-MS	HP-5MS (30 m×0.25 mm; 0.25 μm)	Gradient temperature: 90 °C for 1 min. followed by increases of 5 °C min^{-1} until 290 °C	1 mL min^{-1}	ESI MS	44

Biological fluids	Centrifugation	HPLC	XDB-C8 (4.6×150 mm; 5 μm)	Isocratic elution with 0.05 M orthophosphoric acid : acetonitrile (60 : 40 v/v)	1.5 mL min^{-1}	UV and ESI MS	45
Dry and fresh fruits	Maceration with ethanol	GC	SE-54 (30 m×0.25 mm; 0.25 μm)	Isothermal programme at 250 °C	1 mL min^{-1}	Flame ionisation detection	46
Pericarp and placenta of fresh fruit	MAE	UHPLC	RP-18e (4.6×100 mm) C18 (50×2.1 mm; 1.7 μm)	Gradient elution with A (acidified water with 0.1% acetic acid) and B (acidified methanol with 0.1% acetic acid)	0.4 mL min^{-1} 0.8 mL min^{-1}	UV MS	20
Fresh fruit	UAE with ethanol	HPLC	EVO C18 (150 mm×3 mm; 2.6 μm)	Isocratic elution with 40% acetonitrile and 60% 20 mM aqueous ammonium formate	0.3 mL min^{-1}	ESI MS and UV	47
Vegetable oils	Maceration with methanol	HPLC	C18 (100 mm×2.1 mm; 3.0 μm)	Linear gradient of A (0.1% formic acid) and B (acetonitrile)	250 μL min^{-1}	ESI MS	48
Dried pepper	Maceration with toluene	GC	5% phenyl (30 m×0.25 mm; 0.1 μm)	Gradient temperature: 130 °C for 1 min followed by increases of 5 °C min^{-1} until 325 °C	n.i.	ESI MS Derivatization with BSTFA	49
Seasoning paste	Maceration with acetonitrile	HPLC	C18 (250×4.6 mm)	Isocratic elution with acetonitrile : deionised water : acetic acid (100 : 1000 : 1 v/v/v)	1 mL min^{-1}	UV	50
Paprika	UAE	HPLC	C18	Gradient elution with A (11% aqueous methanol) and B [isopropanol : acetonitrile : methanol (55 : 35 : 10 v/v/v)]	0.6 mL min^{-1}	UV	51
Fresh fruit	Maceration extraction with methanol followed by solid-phase extraction	GC	ZB-5	Gradient temperature 150–250 °C (20 °C min^{-1}), followed by increase from 250 to 280 °C at 2 °C min^{-1}	1.0 mL min^{-1}	MS	52

[a]HPLC = high-performance liquid chromatography; UAE = ultrasound-assisted extraction; MAE = microwave assisted extraction; UHPLC = ultra-HPLC; UV = ultraviolet; ESI = electrospray ionisation; MS = mass spectrometry; LC = liquid chromatography; GC = gas chromatography; n.i. = not indicated; AP-CI = atmospheric pressure chemical ionisation; MTBE = methyl-*tert*-butyl ether; BSTFA = *N*,*O*-bis(trimethylsilyl)trifluoroacetamide.

most commonly used stationary phases (columns) are non-polar, such as those composed from films of: 100% dimethylpolysiloxane; (5%-phenyl)-methylpolysiloxane; 5% phenyl, 95% dimethylpolysiloxane; (5%-phenyl)(1%-vinyl)-methylpolysiloxane; or 5% diphenyl/95% dimethylpolysiloxane.

It might be thought that the selectivity and sensitivity of gas chromatography are lower than in HPLC due to the absence of an interaction with the stationary phase, but this is not true. Some optimised gas chromatography systems, such as the last example in Table 9.4,[52] are capable of differentiating and quantifying isomers of homodihydrocapsaicin. Gas chromatographic analysis of capsaicin can present a working range from 0.025 to 250 mg mL^{-1}, with limits of detection as low as 0.25 ng.[44,52]

9.5 Mass Spectrometry

Spectrophotometric and chromatographic methods alone cannot give data on the molecular structure of metabolites. Identification by these methods is therefore interpretative, based on the chemical class and physicochemical properties. If the unambiguous identification of compounds is required, methods such as mass spectrometry and NMR must be used. Both techniques provide data on the molecular structure, enabling the identification of isomers, as shown by Maokam *et al.*,[52] whose study is cited as an example of isomer analysis through gas chromatography using a mass spectrometric detector.

In mass spectrometry, the molecules in the sample are forcibly ionised to promote the fragmentation of their molecular structure into ions and the mass-to-charge ratios are measured. The molecular mass of each of the ions constituting the original molecule can therefore be obtained. The fragmentation patterns of most naturally occurring chemical classes, such as flavonoids,[53] alkaloids[54] and terpenoids,[55] have been extensively described. In capsaicinoids, the fragmentation pattern is mainly characterised by cleavage of the vanillyl moiety and the dissociation of a methylene from this ion (Figure 9.3).

The determination of capsaicinoids by mass spectrometry can be achieved using different configurations of the equipment, including direct injection, in which the sample is directly inserted in the ionisation source, or chromatography. The ionisation techniques include electrospray ionisation (ESI), atmospheric pressure chemical ionisation (APCI), electron impact (EI) and direct analysis in real time (DART) in negative and positive ionisation modes. The positive mode is applied more often than the negative mode because the capsaicinoids can be easily protonated.[56–58]

9.6 Nuclear Magnetic Resonance

Like mass spectrometry, NMR spectrometry can unambiguously identify molecules. It is based on the excitation of the spin of atoms when exposed to a magnetic field. The NMR techniques most often applied in capsaicinoid analysis use ^{1}H (Figure 9.4) and ^{13}C excitation. Gómez-Calvario *et al.* summarised the expected ^{1}H and ^{13}C NMR spectral signals of 159 capsaicinoids.[59]

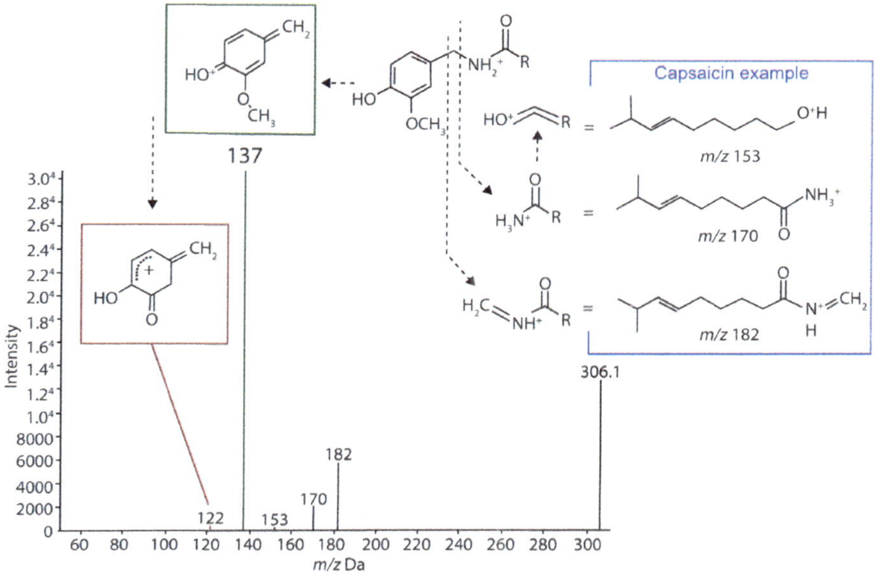

Figure 9.3 Fragmentation pattern of capsaicinoids and the capsaicin fragmentation pattern in positive ionisation mode. Image credit: Carlos Magno da Silva Antonio.

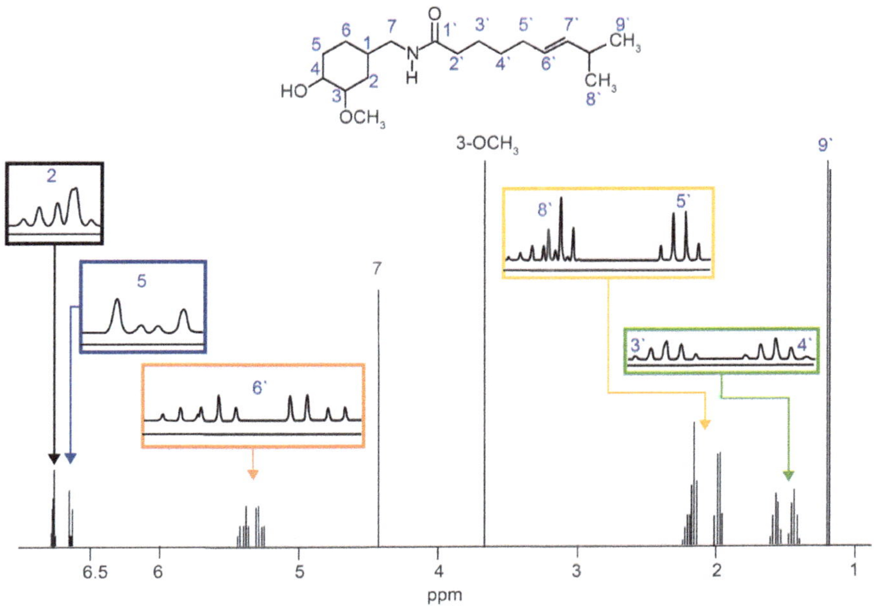

Figure 9.4 Predicted ^{1}H-NMR spectra of capsaicin (400 MHz) (calculation available at https://www.nmrdb.org/13c/index.shtml?v=v2.121.0).[60–62] Image credit: Carlos Magno da Silva Antonio.

Liquid and solid samples of capsaicinoids can be used in NMR analysis. Liquid samples are usually analysed by attaching an NMR spectrometer as the detector in a chromatographic system.[11] Developments in NMR have now enabled the individual quantification of capsaicinoids and their total content without the need for chromatographic separation. These NMR methods offer advantages including a high reproducibility, high sensitivity, a reduced data acquisition time (compared with gas chromatography and HPLC) and less chance of cross-contamination of samples in large sample sets. Chemometric tools can be used to process data to remove the chromatographic separation steps and still obtain reliable data with highly complex samples. These tools include ^{1}H quantitative NMR or magic angle spinning NMR.[63–65] The evaluation of solid samples by NMR can aid in the development of highly sensitive quality control methods for spices and drugs containing capsaicinoids.

9.7 Electrochemical Methods

Mass spectrometry and NMR are remarkable methods as a result of their sensitivity and reliability, but both have relatively high costs of implementation and require highly trained personnel to operate and maintain the equipment and interpret the data. Electrochemical methods avoid these disadvantages, but still give high-quality data. Electrochemical sensors are also more sustainable because they require less material to be built and less solvent to perform the actual analysis. In addition, they can be operated by unskilled personnel and do not always require sample preparation.[66–68]

The determination of capsaicinoids by electrochemical methods is based on oxidation of the vanillyl moiety, which is a proton donor. Vanillyl oxidation can be promoted by electro-oxidation in pH-dependent pathways or by enzymatic oxidation. In the pH range 6–10, vanillyl donates two protons in a two-step electro-oxidation reaction. The hydroxyl hydrogen is donated first, forming a phenoxy ion, and then the hydrogen in $CH\text{-}CHO_3$ is donated to form the phenoxonium cation. At pH <6, the oxidation goes straight to the formation of the phenoxonium cation, releasing two electrons in a single step in a faster reaction (Figure 9.5). The potential generated in both cases is used to determine the capsaicinoids. The pH of the medium is usually <6 to enable faster data acquisition.[11]

Several electrochemical methods are used to determine capsaicinoids and can be applied to a wide range of sample types in solid or liquid form, such as sauces, powdered peppers, paprika preparations and pepper extracts.[11] The working electrodes can be made from various materials and use several different voltammetric measuring methods.[69–71] Examples of working electrodes for the determination of capsaicinoids include: basal pyrolytic graphite with multiwalled carbon nanotubes; a carbon paste with vinyl ferrocene or β-cyclodextrin; a carbon screen-printed electrode with vinyl ferrocene; graphite electrodes; platinum electrodes; and glassy carbon electrodes with

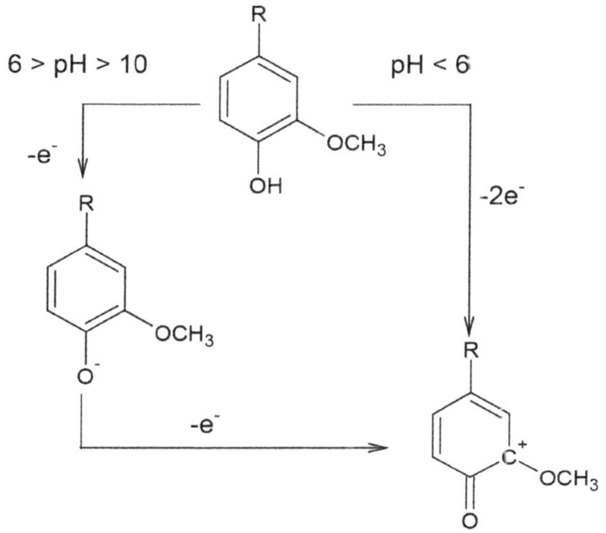

Figure 9.5 Electro-oxidation of capsaicinoids. Image credit: Carlos Magno da Silva Antonio.

multiwalled carbon nanotubes, gold particles, ruthenium nanoparticles or paraffin. The voltammetric measuring methods include: cyclic voltammetry (most applied); linear sweep voltammetry; adsorptive stripping voltammetry; differential pulse voltammetry; and square wave voltammetry.

Such versatility has enabled the development of methods with performance features comparable with HPLC, which is currently considered to be the standard method for the quality control of pepper products. The limit of detection of electrochemical methods can be as low as 0.002 μM, with working ranges up to 138 $\mu mol\,L^{-1}$,[69] enabling broad application options. Despite the advantages of electrochemical methods, attention must be given to interferents in the analysis – for instance, phenolic compounds, vanillin derivates and ascorbic acid can influence the analytical response of these methods.

9.8 Comparison of Methods

Even with its focus on the determination of capsaicinoids, this chapter gives an indication of the large number of options available for the analysis of *Capsicum* species. The most important point is that you should use the method most suitable for your purpose and situation. You can only compare methods that are used for the same purpose. The initial questions that must be addressed are: (1) can part of my sample be destroyed?; (2) do I need quantitative data?; (3) how accurate must the data be?; and (4) is it necessary to determine individual metabolites? All these questions have been addressed in this chapter and are summarised in Table 9.5.

Table 9.5 Comparison of different analytical methods used for *Capsicum*.[a]

Method	Type of data	Analyte	Can sample be recovered?	Estimated working range	Additional features	Major applications
Scoville heat test	Semi-quantitative	Only pungency is determined	No	0–3 000 000 SHU	Low reproducibility	Fruits and sauces
Colorimetry (with or without thin-layer chromatography)	Qualitative	Presence of capsaicinoids	No	—	Toxic reagents, high consumption of solvents, high production of residues, phenolic compounds are interferents	*Capsicum* fruit extracts and oleoresins
High-performance thin-layer chromatography with densitometry	Qualitative, quantitative	Total capsaicinoid content	No	50–1000 μg mL^{-1}		
UV spectrometry with separation methods	Qualitative, quantitative	Total capsaicinoid content; individual content of major capsaicinoids	Yes	1–1000 mg L^{-1}	Simple execution, low cost, but phenolic compounds, alkaloids and carotenoids are interferents	*Capsicum* fruit extracts and oleoresins, enriched fractions of capsaicinoids
Near-infrared spectrometry	Quantitative	Individual content of major capsaicinoids	Yes	0.2–100 mg kg^{-1}	No sample preparation step, large set of data, but flavonoids, carotenoids and water are interferents	Powered dry pepper, solid samples, isolated capsaicinoids, quality control of paprika
High-performance chromatography (HPLC, UHPLC and GC)	Quantitative	Total content; individual content of major and minor capsaicinoids	Depends on equipment	0.001 μg L^{-1} to 100 mg L^{-1}	Requires a tandem detection method, time-consuming analysis, requires skilled analyst, high operational costs	*Capsicum* fruit extracts and oleoresins, enriched fractions, isolated capsaicinoids, bioproduct extracts, drug extracts, biological fluids, spices
Mass spectrometry	Qualitative	Identification of major and minor capsaicinoids	No	—	Can be used as a detector in chromatographic methods, unable to distinguish conformational isomers	
Nuclear magnetic resonance spectrometry	Qualitative, quantitative	Identification of major and minor capsaicinoids; total content of capsaicinoids; individual content of major capsaicinoids	Yes	0.05–5 mg	High instrumental cost, requires skilled analyst, requires chemometric methods, no cross-contamination from sample to sample	Liquid extracts of *Capsicum* and its bioproducts, solid samples of drugs and spices
Electrochemical analysis	Qualitative, quantitative	Total content; individual content of major capsaicinoids	Yes	0.5–138 μg L^{-1}	Rapid execution, low cost, requires development to determine possible interferents	Solid and liquid samples, sauces, powered pepper

[a]HPLC = high-performance liquid chromatography; UHPLC = ultra-HPLC; GC = gas chromatography.

In terms of precision, accuracy and sensitivity, NMR and mass spectrometry are by far the best methods. However, NMR is still developing to provide quantitative data and mass spectrometry alone can only give qualitative results. If we consider that only quantitative data about the four major capsaicinoids are required, then electrochemical methods are evolving and already give an analytical performance comparable with gas chromatography-mass spectrometry. Quantitative NMR and electrochemical sensors are promising methods in the characterisation of *Capsicum* and the quality control of bioproducts, but still require further development of standard procedures to acquire and analyse data.

Despite their simplicity, TLC and spectrophotometric methods are worth using for qualitative data or even to rapidly obtain quantitative data (*e.g.* the total content of capsaicinoids). Preliminary and rapid execution methods are essential in tracking the quality control of large sample sets.

9.9 Concluding Remarks

The key point of this chapter is to show that the choice of analytical method depends on the experimental purpose or situation. If we only need to know the pungency in SHU, any of the methods discussed are suitable. To determine the total capsaicinoid content when monitoring an industrial production line, non-destructible methods with rapid execution are preferred, such as NIR spectrometry and electrochemical methods. However, if the analysis is to be used in a legal process – for instance, in the quality control of a commercial product – then highly sensitive methods that give both qualitative and quantitative data are recommended, such as mass spectrometry and NMR with or without a chromatographic method. When choosing a suitable method, be aware of the limit of detection, the working range, sample compatibility and the sample size.

References

1. J. E. Woodbury, *J. AOAC Int.*, 1980, **63**, 556–558.
2. F. Korel, N. Bağdatlioğlu, M.Ö. Balaban and Y. Hişil, *J. Agric. Food Chem.*, 2002, **50**, 3257–3261.
3. J. Lim, G. Kim, C. Mo and M. Kim, *Sensors*, 2015, **15**, 27420–27435.
4. A. Hayden and C. B. Jordan, *J. Am. Pharm. Assoc.*, 1941, **30**, 107–133.
5. W.-K. Ryu, H.-W. Kim, G.-D. Kim and H.-I. Rhee, *J. Food Drug Anal.*, 2017, **25**, 798–803.
6. P. Dawan, T. Satarpai, P. Tuchinda, J. Shiowatana and A. Siripinyanond, *Talanta*, 2017, **162**, 460–465.
7. B. Estrada, M. A. Bernal, J. Díaz, F. Pomar and F. Merino, *J. Agric. Food Chem.*, 2000, **48**, 6234–6239.
8. N. J. Amruthraj, J. P. Preetam Raj and L. Antoine Lebel, *Int. J. Pharmacol. Clin. Res.*, 2014, **6**, 159–164.

9. K. Lahbib, S. Dabbou, S. E. L Bok, G. Pandino, S. Lombardo and M. E. L. Gazzah, *Ind. Crops Prod.*, 2017, **104**, 164–170.
10. M. Karadžić Banjac, S. Z. Kovačević, Ž. M. Ilin, A. N. Tepić Horecki, B. D. Adamović, A. S. Vakula, Z. M. Šumić and S. O. Podunavac-Kuzmanović, *J. Food Process. Preserv.*, 2020, **44**, 1–8.
11. A. da Silva Antonio, L. S. M. Wiedemann and V. F. da Veiga Junior, *Food Anal. Methods*, 2019, **12**, 1327–1345.
12. D. Y. Kwon, Y. S. Kim, S. Y. Ryu, M. R. Cha, G. H. Yon, H. J. Yang, M. J. Kim, S. Kang and S. Park, *J. Nutr. Biochem.*, 2013, **24**, 1078–1085.
13. A. Das, S. Kundu and B. Ghosh, *Res. J. Pharm., Biol. Chem. Sci.*, 2015, **6**, 606–612.
14. E. Goci, E. Haloçi, K. Vide and L. Malaj, *Albanian J. Pharm. Sci.*, 2014, **1**, 16–19.
15. H. A. A. Gibbs and L. W. O'Garro, *HortScience*, 2004, **39**, 132–135.
16. M. Monforte-González, F. Medina-Lara, G. Gutiérrez-Carbajal and F. Vázquez-Flota, *J. Liq. Chromatogr. Relat. Technol.*, 2007, **30**, 1697–1704.
17. O. Monago-Maraña, M. Guzmán-Becerra, A. Muñoz de la Peña and T. Galeano-Díaz, *J. Food Compos. Anal.*, 2018, **67**, 10–18.
18. A. González-Zamora, E. Sierra-Campos, R. Pérez-Morales, C. Vázquez-Vázquez, M. A. Gallegos-Robles, J. D. López-Martínez and J. L. García-Hernández, *J. Chem.*, 2015, **2015**, 1–10.
19. T. Stipcovich, G. F. Barbero, M. Ferreiro-González, M. Palma and C. G. Barroso, *Food Chem.*, 2018, **239**, 217–224.
20. G. F. Barbero, A. Liazid, M. Ferreiro-González, M. Palma and C. G. Barroso, *Int. J. Food Prop.*, 2016, **19**, 984–992.
21. R. Karnka, M. Rayanakorn, S. Watanesk and Y. Vaneesorn, *Anal. Sci.*, 2002, **18**, 661–665.
22. M. Nag, J. Chanda, R. Biswas, N. A. Al-Dhabi, V. Duraipandiyan, P. Banerji and P. K. Mukherjee, *Indian J. Pharm. Educ. Res.*, 2017, **51**, 337–342.
23. A. Schmidt, G. Fiechter, E. M. Fritz and H. K. Mayer, *J. Food Compos. Anal.*, 2017, **60**, 32–37.
24. J. Lim, C. Mo, S. H. Noh, S. Kang, K. Lee and M. S. Kim, *Int. Soc. Opt. Eng.*, 2012, **8369**, 83690Z–83690Z-8.
25. I. Domínguez-Martínez, O. G. Meza-Márquez, G. Osorio-Revilla, J. Proal-Nájera and T. Gallardo-Velázquez, *J. Korean Soc. Appl. Biol. Chem.*, 2014, **57**, 133–142.
26. G. A. Cordell and O. E. Araujo, *Ann. Pharmacother.*, 1993, **27**, 770–776.
27. B. Horn, S. Esslinger, M. Pfister, C. Fauhl-Hassek and J. Riedl, *Food Chem.*, 2018, **257**, 112–119.
28. I. Reinholds, V. Bartkevics, I. C. J. Silvis, S. M. van Ruth and S. Esslinger, *J. Food Compos. Anal.*, 2015, **44**, 56–72.
29. C. Mo, K. Lee, J. Lim, S. Kang, H. Lee and B. Cho, *Sens. Agric. Food Qual. Saf. III*, 2011, 80270T.
30. J. Jiang, H. Cen, C. Zhang, X. Lyu, H. Weng, H. Xu and Y. He, *Postharvest Biol. Technol.*, 2018, **146**, 147–154.

31. I. Perucka and W. Oleszek, *Food Chem.*, 2000, **71**, 287–291.
32. I. Tolan, D. Ragoobirsingh and E. Y. S. A. Morrison, *Phytother. Res.*, 2001, **15**, 391–394.
33. M. Lal, A. Kadam and A. S. Padole, *J. Drug Delivery Ther.*, 2019, **9**, 169–172.
34. G. S. Subramanian, A. Karthik, S. B. Kamath, K. Prabahar, A. Ranjithkumar, S. Pathak and N. Udupa, *J. Planar Chromatogr.*, 2008, **21**, 6007–6010.
35. C. E. Wagner, T. M. Cahill and P. A. Marshall, *J. Chem. Educ.*, 2011, **88**, 1574–1579.
36. S. Das, K. C. Teja, B. Duary, P. K. Agrawal and S. S. Bhattacharya, *Sci. Hortic.*, 2016, **213**, 354–366.
37. M. R. Maya, I. Ramanaiah, K. Venkatakrishna, K. Rameshkumar, V. Veeramanikandan, M. Eyini and P. Balaji, *J. Drug Delivery Ther.*, 2019, **9**, 485–495.
38. J. C. dos, S. Almeida, T. D. S. Rodrigues, K. F. De Souza, R. G. Rodrigues-das-Dores and T. J. Nagem, *Infarma – Ciênc. Farm.*, 2015, **27**, 106.
39. A. Garcés-Claver, M. S. Arnedo-Andrés, J. Abadía, R. Gil-Ortega and A. Álvarez-Fernández, *J. Agric. Food Chem.*, 2006, **54**, 9303–9311.
40. F. Conforti, G. A. Statti and F. Menichini, *Food Chem.*, 2007, **102**, 1096–1104.
41. J. Pauwels, W. D'Autry, L. Van den Bossche, C. Dewever, M. Forier, S. Vandenwaeyenberg, K. Wolfs, J. Hoogmartens, A. Van Schepdael and E. Adams, *J. Pharm. Biomed. Anal.*, 2012, **60**, 51–58.
42. D. Giuffrida, P. Dugo, G. Torre, C. Bignardi, A. Cavazza, C. Corradini and G. Dugo, *Food Res. Int.*, 2014, **65**, 163–170.
43. M. Sganzerla, J. P. Coutinho, A. M. T. de Melo and H. T. Godoy, *Food Res. Int.*, 2014, **64**, 718–725.
44. S. Saha, S. Walia, A. Kundu, C. Kaur, J. Singh and R. Sisodia, *Int. J. Food Prop.*, 2014, **18**, 1535–1545.
45. M. Kuzma, K. Fodor, G. Maász, P. Avar, G. Mózsik, T. Past, E. Fischer and P. Perjési, *J. Pharm. Biomed. Anal.*, 2015, **103**, 59–66.
46. M. R. Loizzo, A. Pugliese, M. Bonesi, F. Menichini and R. Tundis, *LWT–Food Sci. Technol.*, 2015, **64**, 623–631.
47. L. Duelund and O. G. Mouritsen, *Food Chem.*, 2017, **221**, 913–918.
48. F. Ma, Q. Yang, B. Matthäus, P. Li, Q. Zhang and L. Zhang, *J. Chromatogr. B: Anal. Technol. Biomed. Life Sci.*, 2016, **1021**, 137–144.
49. K. G. Sweat, J. Broatch, C. Borror, K. Hagan and T. M. Cahill, *Food Chem.*, 2016, **210**, 606–612.
50. C. Y. Cheok, B. Sobhi, N. Mohd Adzahan, J. Bakar, R. Abdul Rahman, M. S. Ab Karim and Z. Ghazali, *Food Chem.*, 2017, **216**, 10–18.
51. Z. Nagy, H. Daood, A. Koncsek, H. Molnár and L. Helyes, *J. Liq. Chromatogr. Relat. Technol.*, 2017, **40**, 199–209.
52. C. Maokam, S. Techawongstien and S. Chanthai, *Int. Food Res. J.*, 2014, **21**, 2237–2243.
53. F. Cuyckens and M. Claeys, *J. Mass Spectrom.*, 2004, **39**, 1–15.

54. H. J. Shim, J. Y. Lee, B. Kim and J. Hong, *Mass Spectrom. Lett.*, 2013, **4**, 79–82.
55. G. P. Aguiar, E. J. Crevelin, H. J. Dias, S. R. Ambrósio, J. K. Bastos, V. C. G. Heleno, R. Vessecchi and A. E. M. Crotti, *J. Mass Spectrom.*, 2018, **53**, 1086–1096.
56. R. Q. Thompson, K. W. Phinney, M. J. Welch and E. White V, *Anal. Bioanal. Chem.*, 2005, **381**, 1441–1451.
57. T. Wu, X. Yuan, X. Wu, Y. Tang, H. Lin, Z. Che, W. Zhou and W. Li, *Food Anal. Methods*, 2019, **12**, 32–40.
58. A. Arrizabalaga-Larrañaga, G. Campmajó, J. Saurina, O. Núñez, F. J. Santos and E. Moyano, *LWT*, 2020, **139**, 110533.
59. V. Gómez-Calvario, M. L. Garduño-Ramírez, I. León-Rivera and M. Y. Rios, *Magn. Reson. Chem.*, 2016, **54**, 268–290.
60. D. Banfi and L. Patiny, *Chimia*, 2008, **62**, 280–281.
61. C. Steinbeck, S. Krause and S. Kuhn, *J. Chem. Inf. Comput. Sci.*, 2003, **43**, 1733–1739.
62. A. M. Castillo, L. Patiny and J. Wist, *J. Magn. Reson.*, 2011, **209**, 123–130.
63. P. K. Bora, P. Kemprai, R. Barman, D. Das, A. Nazir, S. P. Saikia, D. Banik and S. Haldar, *Phytochem. Anal.*, 2021, **32**, 91–103.
64. P. Siudem, J. Bukowicki, I. Wawer and K. Paradowska, *RSC Adv.*, 2020, **10**, 18082–18092.
65. R. Velásquez-Valle, N. Villa-Ruano, D. Hidalgo-Martínez, L. G. Zepeda-Vallejo, N. Pérez-Hernández, C. A. Reyes-López, E. Reyes-Cervantes, D. L. Medina-Melchor and E. Becerra-Martínez, *Food Res. Int.*, 2020, **131**, 108863.
66. M. T. Moreno, R. E. Brito, M. Córdoba and J. M. R. Mellado, *J. Chem. Res.*, 2020, 1–5.
67. A. Soleh, K. Saisahas, K. Promsuwan, P. Thavarungkul, P. Kanatharana and W. Limbut, *ACS Appl. Nano Mater.*, 2020, **3**, 10094–10104.
68. P. B. Deroco, O. Fatibello-Filho, F. Arduini and D. Moscone, *Electrochim. Acta*, 2020, **354**, 136628.
69. M. Sarma and M. del Valle, *Electroanalysis*, 2020, **32**, 230–237.
70. M. Wang, B. Gao, Y. Xing and X. Xiong, *Int. J. Electrochem. Sci.*, 2020, **15**, 8437–8449.
71. W. Numphud, O. Chienthavorn and W. Siriwatcharapiboon, *ScienceAsia*, 2020, **46**, 586–594.

CHAPTER 10

Peppers as Non-lethal Weapons

10.1 Introduction

Have you ever eaten an extremely hot chilli pepper? If the answer is yes, then you probably felt your eyes watering, your breath being taking away and an intense burning sensation in your throat. These are some of the effects caused by the acute toxicity of chilli peppers. Most mammals exposed to these effects become temporarily unable to fight or even run. These effects do not only occur when you ingest chilli peppers, but also if you are exposed to a spray containing pepper extracts or oleoresin of *Capsicum* (Chapter 8). As the spray reaches your skin, eyes, nose and mouth, the effects of acute toxicity will be felt immediately.

Despite being extremely painful, pepper sprays are relatively safe and only cause permanent side-effects if used inappropriately, such as being sprayed in excessive quantities against a single person. As a result of their ability to immediately incapacitate a person, pepper sprays (Figure 10.1) are used as riot control agents, in personal defensive devices or as non-lethal weapons (NLWs). Another terminology that can be used is "less-lethal weapons", as deaths have occurred in sensitive people or those who have cardiopathic comorbidities.[1]

Most riot control agents are chemical incapacitants, which, by definition, immediately and temporarily incapacitate a person by causing an irritant effect in the eyes, skin and respiratory tract. Riot control agents include synthetic compounds – such as 2-chlorobenzylidene malononitrile (CS gas), 1-chloroacetophenone (CN gas) and dibenzoxazepine (CR gas) – and natural bioproducts, such as the capsaicinoids of *Capsicum* species, the gingerol of *Zingiber* species and the piperine of *Piper* species.[2]

Food Chemistry, Function and Analysis No. 37
Chemistry and Nutritional Effects of Capsicum
By Valdir Florencio da Veiga, Jr, Larissa Silveira Moreira Wiedemann,
Claudio Pereira de Araujo, Jr and Ananda da Silva Antonio

Published by the Royal Society of Chemistry, www.rsc.org

Figure 10.1 Pepper spray device (*spray de pimenta* in Portuguese). Reproduced with permission from Condor Non-Lethal Technologies.

Most of the products were developed or consolidated as NLWs during the First World War and the Cold War. NLWs were used to force enemies out of hiding places, such as caves, underground facilities or trenches. Until the 1990s, the most used NLWs were synthetic compounds. The use of these, however, was discouraged due to their high toxicity and prolonged effects, which led to several deaths, their use as instruments of torture and the view that they violated human rights when used against civilian crowds. In addition, synthetic NLWs are highly inflammable and can cause fires when used in closed areas.[3,4] The use of synthetic NLWs is currently restricted to the military forces and they have mostly been replaced by deterrent weapons based on the oleoresin of *Capsicum*, the so-called pepper sprays.[2] Because peppers sprays have few side-effects and a negligible mortality rate, their use is permitted in several countries as personal defence devices or as an NLW for private protection forces and the police.

10.2 Historical Applications of *Capsicum* as a Weapon

The use of chilli peppers as a weapon is not restricted to modern societies and dates back to the fourth century BC. Humans not only used plants for food and to heal diseases; there are records of ancient communities in China and Brazil that burned peppers and used their smoke against predators and other communities because both animals and humans are blinded and disoriented by their pungent smoke.

In the pre-Columbian era, several indigenous civilisations used burning pepper plants to stop invaders. The indigenous Tupinambás tribe of Brazil used peppers in defensive lines against the Portuguese colonisers. The Aztec civilisations not only prized chilli peppers for their tchocoatl drinks, but also as an offensive weapon. There are records of a warrior squad of Moctezuma II being ambushed and killed by asphyxiation with burning chilli pepper plants in the town of Cuetlaxtlan as its lords were trying to avoid paying tributes.

Even after the start of the Age of Discovery, chilli peppers were still being used as weapons. During the Tokugawa shogunate in Japan, some warriors and police used a weapon call *metsubushi*, which means "eye closers". A *metsubushi* was a small ball filled with pepper powder that was hurled in the enemies' eyes, or burned and used as a grenade.

The current pepper spray formulations were created at the end of the 1970s for use in hunting, but it was only in 1987 that the US FBI agent Thomas Ward studied their application as a stunning device against humans. Ward proved their efficacy in controlling offenders without the use of physical combat and several police officers in the USA started to use the sprays. In 1996, the use of pepper sprays as an incapacitating device was widespread, not because of official police usage, but because Ward created an illegal market by being the only producer of the preparation. In 1993, the International Convention on Chemical Weapons banned the use of pepper sprays and capsaicin itself as weapons of warfare, despite their low toxicity. A few years later, in 1997, the Organization for the Prohibition of Chemical Weapons (OPCW) released a statement to its associated countries forbidding the use of several chemical devices in war. This statement also prohibited the use of pepper sprays with capsaicin in warfare and created a regulation for their use as a domestic riot control agent. Security forces (public or private) and civilians could use pepper sprays in self-defence following local laws.[2,5]

In the last few decades, personal defence devices have become more popular as they have been commercialised for civilians. Although pepper sprays are currently legal in several countries and are considered to be less-lethal weapons, there have been a number of incidents since their popularisation. Toprak *et al.*[1] reported seven deaths due to conflicts involving the use of pepper sprays. In most cases, the victims had previous cardiovascular disease or were under the influence of hallucinogenic drugs before the confrontation. One of these cases shows how the inappropriate use of a device considered to be safe can cause casualties. This case was of a citizen who had a disagreement in a restaurant with an off-duty police officer. During the argument, the officer used the pepper spray in the face of the citizen at least 12 times. The intense exposure caused oedema, airway obstruction, bronchitis and acute aspiration of their gastric contents, which resulted in asphyxia and death 22 minutes after exposure.[1] Between 1990 and 2015, there were 7156 recorded cases of severe injuries with pepper spray across 11 countries, including the USA, India, France and the UK.[6,7]

10.3 Pepper Sprays

Always remember that commercial pepper sprays are non-lethal devices and the reported fatal incidents were due to their misuse. Pepper sprays are devices for long- to short-range defensive situations against a single assaulter or a crowd. The best practice is to use the spray in a single shot with a duration of 3–5 s, aiming at the aggressor's eyes (Figure 10.2). With such practice, the risk of injuries and accidental mortality is drastically reduced and it can provide the necessary time to immobilise the aggressor or escape the dangerous situation. It is also preferable that the user maintains a distance of at least 1 m from the aggressor to prevent exposure to the pepper spray.

Industrially produced pepper sprays use the oleoresin of *Capsicum* as the active compound. Civilians commonly think that pepper sprays and tear gases are the same, but tear gases are produced with CS and CN as the active compounds. The market for pepper sprays is dominated by Mace Brand, Sabre, UDAP Industries, Safariland, Fox Labs International, Zarc International and Condor Non-Lethal Technologies, although there are several other smaller legal brands on the market. The pepper spray market is expected to grow in the USA by 5.7% CAGR (compound annual growth rate) per year from 2015 to 2025. The CAGR is a metric measure of the rentability of a market, which indirectly measures its growth and commercial value. For instance, the CAGR of tissue paper (including toilet tissue) is 5% per year for the time period 2019–2027.[8]

The preparation of commercial pepper sprays includes three major constituents: the oleoresin of *Capsicum*, a diluent and a propellant (Figure 10.3). The exact composition varies between brands. During production, the oleoresin of *Capsicum* is diluted to guarantee that it remains as a liquid within the can and maintains the legal pungency level of the product. The mixture is

Figure 10.2 Demonstration of the use of a pepper spray device. Reproduced with permission from Condor Non-Lethal Technologies.

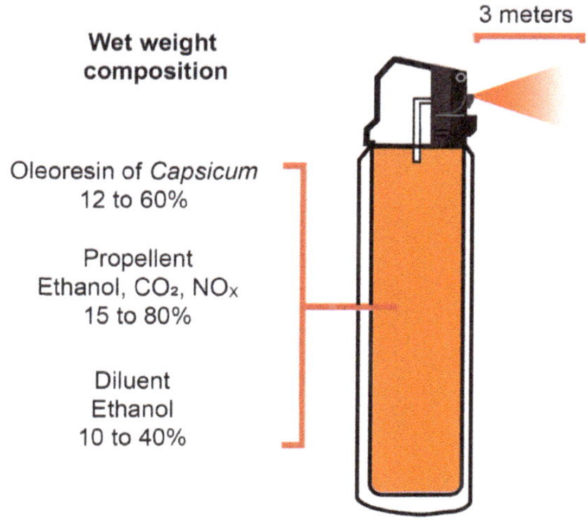

Figure 10.3 General features of a commercial pepper spray for civilian use. Image credit: Carlos Magno da Silva Antonio.

then pressurised with the propellant, which will enable it to be used as a spray with a cone or funnel format, producing small, particle-sized droplets.

Commercial pepper brands for civilians have distinct sizes and can formats. The most common sizes contain 15 mL and last for 25 quick sprays. The cans are usually cylindrical or in gun form, enabling the use of reload units (Figure 10.4). The gun format is more suitable for long-range situations and is only used by security agents and for home protection. The range of a gun spray is twice that of cylindrical cans. If you intend to obtain a self-defence pepper spray, be sure to use an official product from a producer that abides by guidelines or has quality control practices. Some civilians still produce home-made pepper sprays, which is illegal due to their unpredictable pungency. The use of home-made pepper sprays has already led to criminal activities. In 2017 and 2018, home-made pepper sprays were used in crowded places in Italy to cause enough chaos to rob people. In both cases, dozens of people were injured and some fatalities occurred during the following rush and disorder of victims.[5]

Attacks such as those that took place in Italy made pepper sprays less popular. Civilians are usually afraid of the use of pepper sprays, although the main concept in the modern development of non-lethal devices is the use of proportional force to restrain aggression. They have been developed to contain a situation, not to cause damage. Several companies and producers, such as Condor Non-Lethal Technologies, provide specific training on how to use their products and are obtaining successful results. Within this training and the company's philosophy, it is always suggested to first provide verbal warnings before using a deterrent device. During the 2013 Confederations Cup riots in Brazil, Condor pepper sprays were extensively used by the national forces to stop acts of vandalism and violent protest and there were no severe injuries or deaths.

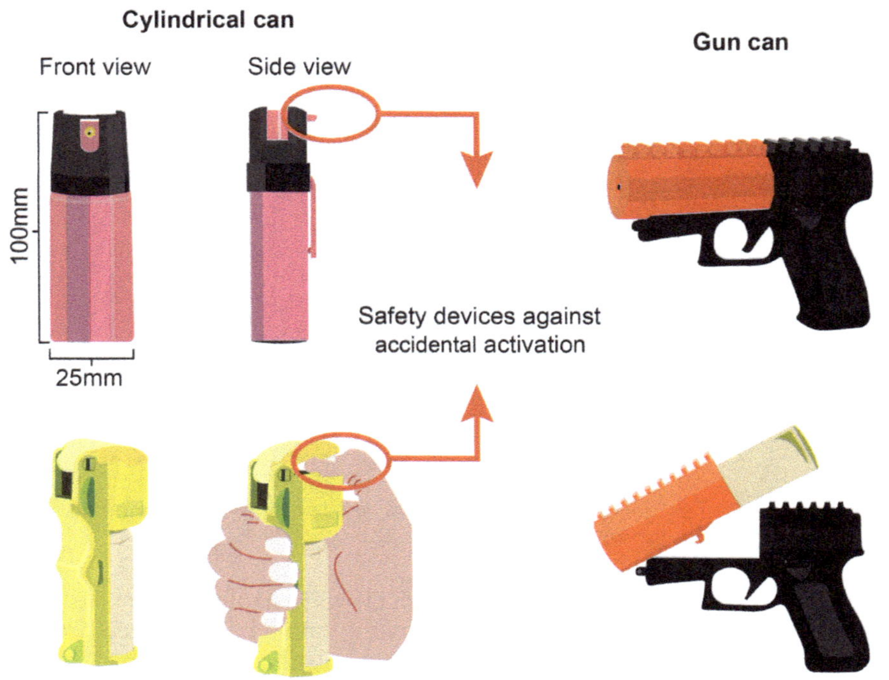

Figure 10.4 Commercial presentation of a pepper spray. Image credit: Carlos Magno da Silva Antonio.

10.4 Legal Issues

Despite the intervention of the OPCW, it is legal in some countries to carry a concealed commercial pepper spray; in fact, there is no universal statement for members of the OPCW to follow. Even after several cases of injuries and casualties due to the irresponsible use of pepper sprays, the records of such cases are low in relation to their high usage. With few reports of injuries related to pepper sprays, little effort has been made to regulate these devices.[1,5,6] Consequently, little attention has been given to health issues, such as toxic effects, and the pathophysiology. Such knowledge is mandatory in the construction of regulatory statements. In addition, the current lack of scientific research on the use of pepper sprays impedes technological developments, the creation of more efficient formulations and the design of quality control methods. There are few published papers on the side-effects of pepper sprays. Some studies were carried out after specific events, such as the home-made grenade attacks in Italy.

Regulation across continents and countries is heterogenous and difficult to assess as the classification of pepper sprays as self-defence devices or offensive weapons varies between regions (Figure 10.5). For instance, pepper sprays are considered as self-defence devices in the USA and can be used by any civilian with a minimum age and no criminal record. In Australia,

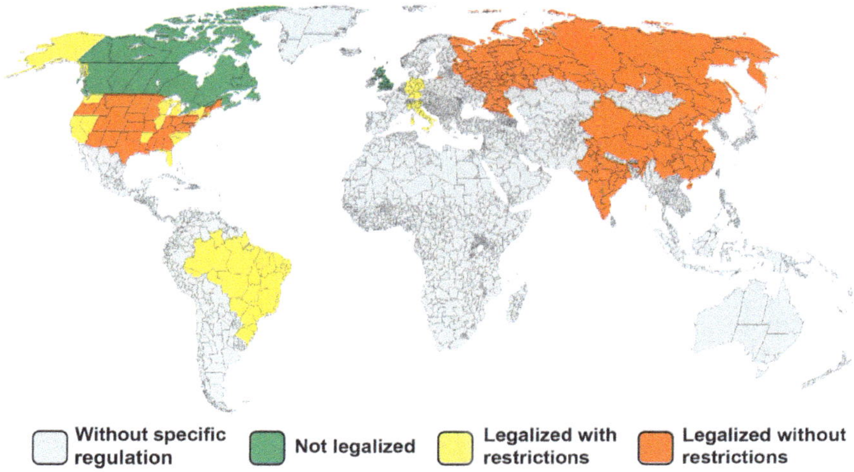

Figure 10.5 Countries with legalisation regarding the commercialisation of pepper sprays to civilians. Image credit: Carlos Magno da Silva Antonio.

pepper sprays are considered as offensive weapons and are prohibited to civilians without lawful reason or when there is imminent threat to life (common in rural areas). There are social movements in Australia to legalise pepper spray as self-defence devices for women due to the increase in violence towards women.

In Europe, some countries (*e.g.* Austria and the Czech Republic) classify pepper sprays as self-defence devices that can be concealed when carried. In other countries, such as Iceland and the UK, they are offensive weapons. In Italy, similar to the constraint in the USA, there are restrictions regarding the minimum age, as well as volume of the device and the formulation, in which the capsaicinoid content cannot exceed 2.5% of the total formula weight.[2]

10.5 Pepper Spray Toxicity

The toxicity and pharmacology of pepper sprays have been studied since the 1920s. The deterrent effect of pepper sprays is related to the capsaicinoids, mainly capsaicin. Capsaicin induces pain because it interacts with the transient receptor potential vanilloid 1 (TRPV1) cation channels in the nerves of several mucous membranes in the face, including the eyes, nose, mouth and throat. For this reason, pepper sprays are more effective when aimed at the face of the aggressor. The stimulation of these nerves causes a sensation of heat and induces the body to protect itself against burning. TRPV1 is also responsible for inducing an inflammatory response in the body and its activation leads to severe, acute pain.[7,9]

The physical consequences of TRPV1 stimulation are lacrimation, the involuntary closure of the eyes, coughing, coryza, apnoea, difficulty in breathing and a burning sensation in the eyes, skin, nose, mouth and throat

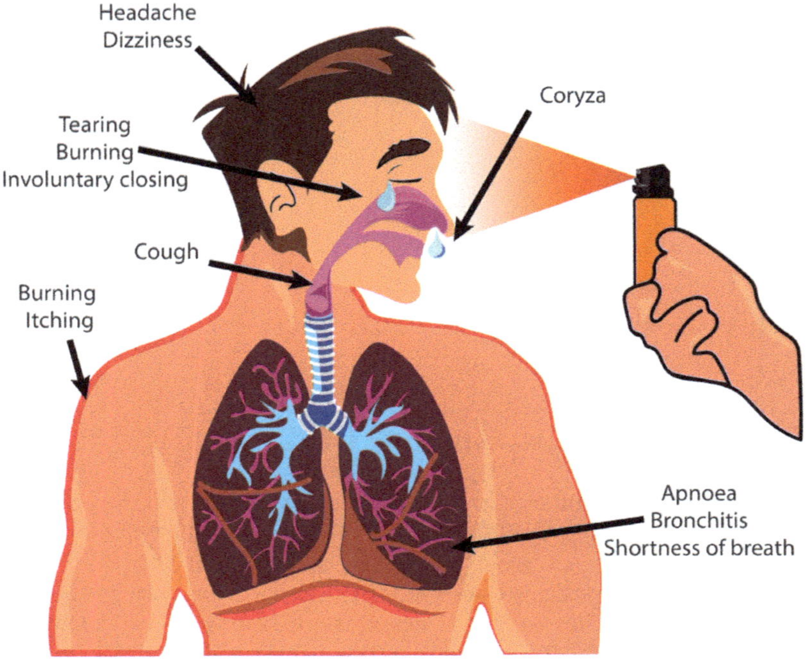

Figure 10.6 Short-term effects of pepper sprays on the aggressor. Image credit: Carlos Magno da Silva Antonio.

(Figure 10.6). Other uncommon symptoms usually associated with long exposure include conjunctivitis and abrasion/erosion/ulceration of the cornea. The eyes and respiratory functions must be closely evaluated after exposure to pepper spray because they are the most sensitive targets of these devices.[2,6]

If a person is hit by a pepper spray, the first aid measures are to contain without becoming contaminated. Touching a person exposed to pepper spray can transmit solid particles of the spray, which may cause pain in bystanders. Material moistened with water, such as a towel or a T-shirt, can be preliminarily used to remove residues in the exposed person. Washing the eyes and face with copious amounts of water is recommended, although, because capsaicin is not very soluble in water, the use of other liquids, such as milk or baby shampoo if available, help to reduce the pain more quickly.[9] The symptoms caused by pepper sprays usually continue for 10–60 minutes if no treatment is provided, if the pepper spray was used correctly and if the person was exposed to just a single, short-duration spray. Exposure is usually more severe when pepper sprays are used during riot control than in a single conflict. In this case, it is recommended that extra precautions and medical attention is given if symptoms are prolonged.

10.6 Quality Control

There is currently no standard, globally recognised method to produce pepper sprays and each commercial brand has its own formulation. Most formulations use oleoresin of *Capsicum* as the source of capsaicinoids because pure capsaicin is extremely toxic. The oleoresin of *Capsicum* used is usually of food grade, the same product commercialised for the food industry. Food-grade oleoresin of *Capsicum* cannot have a high content of capsaicinoids (no higher than 2.5%),[10] but pepper sprays need the capsaicinoids to be effective, so how can we guarantee that the commercial product is effective for its purpose? The development of methods to extract oleoresin of *Capsicum* focused on the production of non-lethal devices is scarce and usually protected by patents.

The methods used specifically to extract oleoresin of *Capsicum* for NLWs mainly use maceration and ultrasound-assisted extraction due to their low cost and execution time.[2,5,11] The content of capsaicinoids in fruits varies as a result of several different factors, with the same variability observed in the oleoresin. It is therefore not possible to guarantee the amount of the active principle, the capsaicinoids, in pepper sprays without proper technological developments, laws, the supply of a standard oleoresin of *Capsicum* and the determination of quality control markers. Oleoresin of *Capsicum* is an extremely complex matrix (see Chapter 8) and it is difficult to control the efficacy and reproducibility of the final product. A proposal that requires further evaluation of toxicity is the application of "natural capsaicin" extracted from the oleoresin of *Capsicum* rather pure capsaicin. Natural capsaicin is composed of 68.7% capsaicin and 31.3% dihydrocapsaicin and is easier to control.[7]

In general terms, the oleoresin of *Capsicum* used in pepper sprays is extracted with ethanol or water and the commercial formulation contains a propellant and a diluting solvent. Pepper sprays commercialised for civilians can have a concentration of oleoresin of *Capsicum* varying from 12 to 60% of the wet weight.[12,13] The capsaicin content within the formulation must vary between 0.90 and 1.60% to ensure that the device will produce an incapacitant effect.[2,14,15]

The propellant and the diluent must have low flammability, toxicity and vapour pressure to ensure safety and evaporation after contact with the intended recipient. The propellants typically used are ethanol, carbon dioxide, nitrogen oxide, propylene glycol dicaprylate/dicaprate and trisglycerol-2-(ethyl hexanoate), which are also used in cosmetic formulations.[2] Because capsaicinoids are more soluble in organic protic solvents, the diluent is usually ethanol because its toxicity is significantly lower than that of other solvents capable of diluting capsaicinoids.

An important concern in pepper spray devices is that their formulations are continually being developed to increase safety. A notable case is one of the Condor products (product reference GL-108/E). This product, provided to law enforcement and police forces, was formulated with natural capsaicin to be applied not as an aerosol jet, but as a foam, similar to a shaving cream (Figure 10.7). Such a formulation has a lower toxicity than CS gas and synthetic capsaicin and it is easier to ensure reproducible concentrations during manufacturing. In addition, the foam-like presentation enhances the accuracy of the user's aim, enabling the selective application of the device in crowds and reducing self-contamination of the user.

Figure 10.7 Demonstration of the use and range of a foam-based pepper spray. Reproduced with permission from Condor Non-Lethal Technologies.

10.7 Concluding Remarks

Capsicum oleoresin has several applications in our daily life, but its use as a defensive device is still controversial and lacks regulation. Is it safe enough to be commercialised to civilians? It can be, with proper scientific investigation, education and clarification of self-defence rights in increasingly violent societies. Pepper sprays for civilians still use an overly complex and variable matrix as the deterrent agent. There have been few investigations of the symptoms of victims and the side-effects of the devices. Further information is required to determine which aspects should be improved to guarantee safety to all affected parties.

References

1. S. Toprak, G. Ersoy, J. Hart and P. Clevestig, *J. Forensic Leg. Med.*, 2015, **29**, 36–42.
2. A. S. Antonio, G. D. Fonseca, Á. J. B. Mendes, L. S. M. Wiedemann and V. F. Veiga-Junior, *J. Braz. Chem. Soc.*, 2019, **30**, 1115–1129.
3. C. J. Hilmas, in *Handbook of Toxicology of Chemical Warfare Agents*, ed. R. Gupta, Academic Press, London, 2nd edn, 2015, vol. 11, pp. 131–150.
4. J. Smith and I. Greaves, *J. Trauma: Inj., Infect., Crit. Care*, 2002, **52**, 595–600.
5. L. Mörén, S. Jonsson, T. Tengel and A. Östin, *Forensic Sci. Int.*, 2019, **304**, 109956.
6. R. J. Haar, V. Iacopino, N. Ranadive, S. D. Weiser and M. Dandu, *BMC Public Health*, 2017, **17**, 1–14.
7. N. Oliveira Junior, V. Padulla, I. Flores, D. Annunciação, M. Neiva and V. Brant, *Hum. Factors Mech. Eng. Def. Saf.*, 2019, **3**, 1–10.

8. Grand View Research, https://www.grandviewresearch.com/industry-analysis/self-defense-products-market, accessed January 2021.
9. R. D. Tidwell and B. K. Wills, *StatPearls [Internet]*, Treasure Island (FL), StatPearls Publishing, 2021. Available from: https://www.ncbi.nlm.nih.gov/books/NBK544263/, acessed Octuber 2020.
10. Food and Agriculture Organization of the United Nations, Prepared at the 35th Expert Committee on Food Additives meeting (1989), published in Compendium of Food Additive Specifications 49 (1990).
11. N. Davison and N. Lewer, Research Report No. 7. Bradford: University of Bradford, Department of Peace Studies, 2005. https://bradscholars.brad.ac.uk/handle/10454/3999, accessed January 2021.
12. W. Bahary, US 2003/0170180 A1, 2003.
13. J. B. D. Pinkney, US 005217708A, 1993.
14. P. M. Blumberg and L. V. Pearce, US 9277748 B2, 2016.
15. K. Loghman-Adham, US 7270802B2, 2002.

CHAPTER 11

Commercial Applications of Capsicum

11.1 Introduction

The genus *Capsicum* undoubtedly contains some of the most important crops worldwide. Despite the origin of the genus in Central and South America, a huge diversity of cultivars is cropped all over our planet. Even before the Age of Discovery, *Capsicum* was commercialised, domesticated, eaten and used as a defensive weapon and its discovery had shaped Indian and Mexican cultures. *Capsicum* peppers are now grown everywhere and are actively present in several cultures and in day-to-day commerce (Figure 11.1).

The unique characteristics of *Capsicum* fruits, from their sensorial attributes to their ability to adapt to different climates and easy domestication, highlighted them as a valuable item of trade. The taste of peppers was unique, the plants were colourful and beautiful, and domestication allowed them to be commercialised at lower prices than the important spices of the Silk Route. It did not take long for *Capsicum* peppers became as popular as black pepper (*Piper nigrum* L.). All social classes were able to afford *Capsicum* and its trade shaped marine routes. *Capsicum* changed and established the global economy as we know it today.

Despite the great attention given to its culinary applications, this peculiar genus is not commercialised just for its taste. The *Capsicum* genus has had different applications throughout human history, reinventing itself as a bioproduct, from ornamental plants to an indispensable food additive and medicines. There are hundreds of bioproducts available that use *Capsicum*

Food Chemistry, Function and Analysis No. 37
Chemistry and Nutritional Effects of Capsicum
By Valdir Florencio da Veiga, Jr, Larissa Silveira Moreira Wiedemann,
Claudio Pereira de Araujo, Jr and Ananda da Silva Antonio

Published by the Royal Society of Chemistry, www.rsc.org

Figure 11.1 An array of different *Capsicum* products in a market. Image credit: Carlos Magno da Silva Antonio.

as one of their ingredients in the form of dry powders, oleoresins, essential oils or extracts.

The *Capsicum* genus has gained increasing attention since the beginning of the twenty-first century, particularly for use as non-edible products. The remarkable biological activity of the genus and its capsaicinoids has stimulated the production of analgesic and anti-inflammatory agents and skin cosmetics. Even more complex health conditions, such as obesity, can use *Capsicum* bioproducts in their treatment. This chapter provides a global perspective of the commercial importance of *Capsicum* and delves into the diversity of bioproducts available.

11.2 *Capsicum* Market

Capsicum has been part of human history for at least 7000 years. Its popularity has changed the world economy since the Age of Discovery and *Capsicum* species remain to this day one of the most prized crops on the planet and are the raw material in several bioproducts. It is estimated that the world trade of *Capsicum* in 2018 moved around 5.44 billion dollars.[1] From 1960 to 2019, the production of *Capsicum in natura* and as a dry fruit continued to increase, reaching 38 027 164 and 4 255 050 tonnes, respectively (Figure 11.2). The expansion of production worldwide in this period is estimated to be 15% for fresh peppers and 23% for dry peppers.[2] Increases in crop yields were even higher at 35 and 46% for dry and fresh peppers, respectively. The production of dry *Capsicum* peppers has increased faster than

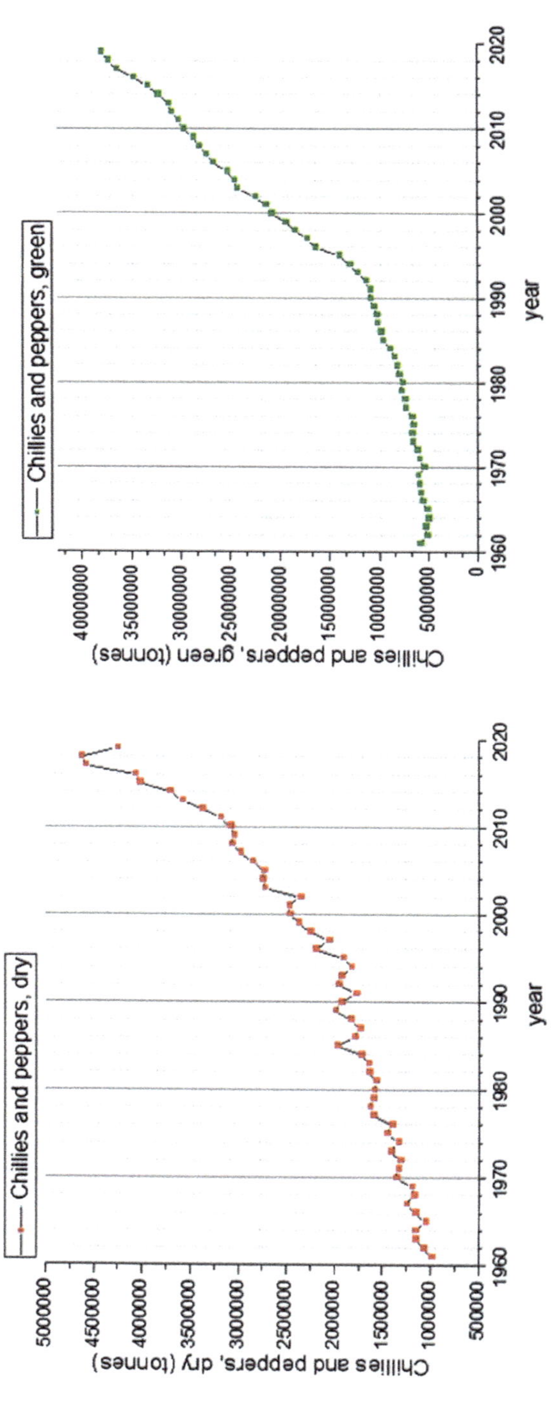

Figure 11.2 Increase in the production of fresh and dried *Capsicum* worldwide since 1960. Image credit: Carlos Magno da Silva Antonio.

that of the fresh fruit (Figure 11.2) as the production of food additives containing *Capsicum*, such as paprika, has become more popular.

According to the Food and Agriculture Organization (FAO) of the United Nations, the major producers of fresh *Capsicum* fruits in 2019 were China and Mexico, producing 18 978 027 and 3 238 245 tonnes, respectively (Figure 11.3). India was the major producer of dried peppers in 2019 with 1 743 000 tonnes, followed by Thailand, China and Ethiopia, with production ranging from 350 000 to 300 000 tonnes. Despite the cultural importance of peppers in Mexico, its production of dry pepper was only 60 706 tonnes. The FAO data do not include Brazil, one of the centres of *Capsicum* diversity. It was estimated that the annual production of fresh fruit in Brazil is 75 000 tonnes.[3,4]

Regarding imports and exports, India, China, Mexico and the USA are the major leaders in the global market and the consumption of *Capsicum* products, although only Mexico is a natural centre of biodiversity. The USA is the leading importer of fresh and dry *Capsicum* peppers (1.78 billion and 321 million dollars, respectively). As one of the major producers of dry pepper, India is also the leading exporter with an 854 million dollar trade value. China holds second place in the export and import of dry pepper. Spain, despite being the fifth major producer, was ranked as the second major exporter of pepper in 2019, with a trade value of 1.35 billion dollars.[1,2]

The market prices of peppers vary according to the bioproduct (*e.g.* dry pepper, fresh fruit, paprika and cultivar) and the season. In the USA, the market price of paprika with a 120 ASTA colour (Chapter 5) is, on average, US$3.80 $\pm$ 0.44 per kg. Fresh pepper export prices have a similar average, with values of US$3.48 and 2.09 per kg during the winter and summer, respectively.[6] Paprika oleoresin can be purchased at US$8.89 per kg depending on its quality. Pepper bioproduct prices vary significantly with the season. In general, the prices are higher when supply is lower or demand is high. Depending on the species and cultivars, this will occur in a specific time period.

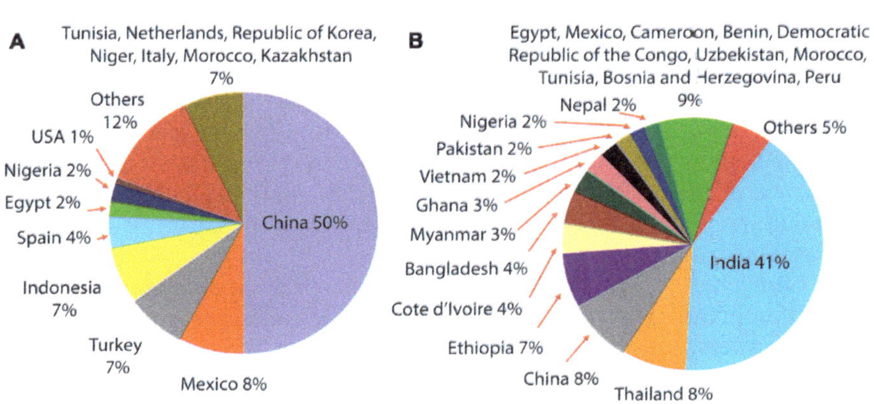

Figure 11.3 Major producers of (A) fresh *Capsicum* fruits and (B) dry *Capsicum*. Image credit: Carlos Magno da Silva Antonio.

For instance, Brazilian varieties such as dedo de moça, malagueta and de cheiro have a higher price from June to September during the winter season in this country. In Indonesia, the fourth largest producer of *Capsicum*, the lowest prices are from April to August and the highest in December when the demand for Christmas celebrations increases during the peak rainy season, which reduces the harvest of chilli peppers.[5]

11.3 Pepper Products

Capsicum species were the seventh most produced fruit worldwide in 2019. The fresh fruits and dry powders of *Capsicum* can be used in the production of several products in different industrial sectors. *Capsicum* can be used to produce spices (*e.g.* paprika and peppers in brine), edible additives (including hot sauces, marinades, candies and chips) and capsules for the wellness industry. In addition, *Capsicum* oleoresin can be used in the production of self-defence weapons, topical creams for pharmaceutical purposes, and other cosmetic and edible products.

11.3.1 Peppers *In Natura*

The fresh fruits of *Capsicum* are mainly used in cooking, as a vegetable, to compose dishes. Several cultures include fresh peppers in their cooking, including Korean, Mexican, Chinese and Indian cuisines. Famous cultivars consumed *in natura* include the sweet bell pepper, habanero, serrano, chipotle, cayenne, jalapeño, ancho, poblano and aji.

One disadvantage in the commercialisation of fresh peppers is that they are very sensitive to manipulation and can be easily crushed during harvesting and storage, reducing their quality and monetary value. If they are not chilled, fresh fruits last for three to seven days post-harvest. As most farms are small in size, their final consumer is usually regional. Transport over long distances can be achieved with refrigeration. The recommended temperature for whole fruits is 7.5 °C with a relative humidity of 95% for a maximum period of three weeks to avoid cold damage and dehydration.

Other methods used to increase the durability of fresh peppers include pre-treatment with pressurised argon before storage and modified atmosphere packaging with low O_2 (2–5%) and CO_2 (5%). *Capsicum* peppers *in natura* are sold in both wholesale and retail markets, but specific statistical data are scarce as a result of the difficulties in monitoring small- and medium-sized suppliers of local markets.

The value chain of *Capsicum* can be extremely branched depending on the region of cultivation (Figure 11.4). It can include consolidators, traders and market retailers, which are intermediaries between the producers and the final consumer. In general, these bodies are responsible for the selection, packing and transportation of the products. Local production, which can make its way directly to the final consumer, has major difficulties due to the lack of experience of some farmers (reducing cultivation yield and quality),

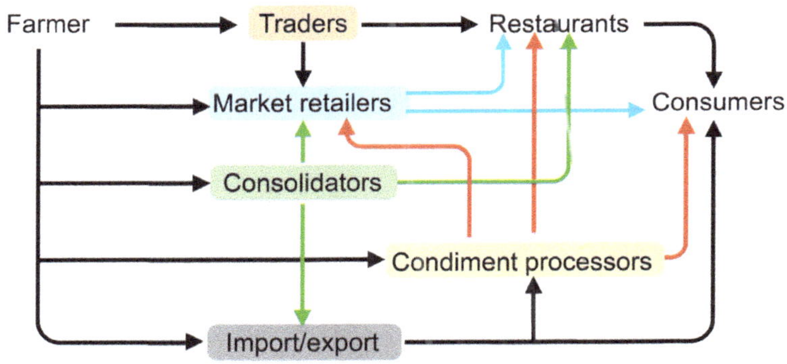

Figure 11.4 Summary of the value chain of *Capsicum in natura*. Image credit: Carlos Magno da Silva Antonio.

an absence of suppliers of native seeds and the low participation of small-scale farmers in larger value chains.

11.3.2 Natural Colourants: Paprika

"A feast for the eyes". This is the first assumption in the use of food colouring additives. Food is primarily consumed *via* the visual sense. The appearance, safety, acceptability and sensory characteristics of foods are all affected by colour. However, not all raw ingredients will have an attractive colour. For instance, if we cook meat with no colourant, it will look pale and unappealing. Because the human perception of flavour is composed of a mixture of sensorial features, including colour, a pale meat will seem to have less taste than a coloured meat, even when the same spices are used on both.

Food colourings can be natural or synthetic. Synthetic colourings currently dominate the food industry as they have low production costs, can be produced faster and have sensorial properties such as aroma that can be easily controlled.[6] In addition to these advantages, synthetic colourings have potential issues and are more restricted by regulatory agencies because they can contain toxic residues, such as copper sulphate, lead chromate and arsenic.[7]

Paprika is one of the major natural food colourings (Figure 11.5). Paprika is the main commercial form of *Capsicum* chilli peppers used as a natural food additive. It is defined as a ground, dry pepper, usually non-pungent, with a bright red colour. The major species used to produce non-pungent paprika is *Capsicum annuum* and the colour can vary in intensity depending on its application. Its production is simple and can be carried out by even small producers. The fruits are cleaned, sliced into small pieces, partitioned from the seeds and left to dry. When completely dry, the pieces are ground into a fine powder. During this process, seeds can be ground together to add pungency to the final product.[8]

Always remember that the red colour of paprika is not related to its pungency, but to its content of red fraction carotenoids (Chapter 4).

Figure 11.5 Jar of commercial paprika. Image credit: Carlos Magno da Silva Antonio.

Any *Capsicum* bioproducts intended to enhance colour or taste are classified as a spice. *Capsicum* consumption as a spice is broad and can vary from 2.5 to 25 g per person per day.[9–11]

One of the key factors during paprika production is the drying method. If the fruit is not dried correctly, then it can easily perish due to microbial proliferation. The optimum moisture content in paprika is 8% to avoid the development of fungal infections. Traditionally, the drying process is called "cure" and can include drying in the sun, drying with hot air (oven-drying) or drying with microwave or infrared radiation.[12] The difference between these methods is the temperature and duration. For instance, sun-drying can take from three days to four weeks under temperatures close to 30 °C, whereas drying in hot air can take from 7 to 30 h at temperatures from 60 to 80 °C.[12,13]

Longer exposure to high temperatures affects the quality of paprika because it is usually evaluated by the content of carotenoids and vitamins, which are thermosensitive. Parameters such as colour retention and the total carotenoid and capsaicinoid content are higher in sun-dried samples than in oven- and infrared-dried samples.[12,13] Microwave drying has been reported[14,15] to produce paprika with a comparable or better quality than the sun-dried product and may contain higher amounts of vitamins, increasing the nutritional value.

Paprika is produced in several different countries, but consumers usually prefer those of Hungarian and Spanish origin. Paprika is a symbol of traditional Hungarian cuisine and it is estimated that the consumption of paprika powder in Hungary is 1.3 g per day.

Hungarian paprika is produced from a mix of dry, powdered and smoked peppers and is prepared in eight grades: (1) special = bright red, very little pungency to non-pungent and very aromatic; (2) capsaicin-free = bright red, non-pungent or some bitterness with a sweet aroma; (3) delicatesse = bright red, non-pungent; (4) hot table = bright red, with at least 25 mg of capsaicinoids per 100 g; (5) fine sweet = yellowish red, little pungency and aromatic; (6) semi-sweet = yellowish red, little pungency; (7) rose (pink) = dull red to pale yellow, pungent; and (8) pungent = dark red, highly pungent.

Spanish paprika is available in three grades (mild, mild spicy and spicy) based on pungency and colour. Hungarian paprika usually has a moisture content of 10%, whereas Spanish paprika has a standard moisture content of 8.0%.[8,16] Despite the cultural importance of paprika in Hungary, the country does not hold an important position in paprika commerce, neither for imports or exports. This is due to two "paprika scandals" in this country, the first in 1994 when paprika was diluted with lead oxide and the second in 2004 when mixing Hungarian paprika with imported paprika resulted in the present of aflatoxins in several products.[16]

The quality of paprika can be evaluated by colour, taste, moisture, total ash, aroma, and the content of capsaicinoids, carotenoids and ascorbic acid. Product safety also requires chemical and microbiological testing.[8] Chemical safety refers mainly to the absence of heavy metals, pesticide residues and adulterants. Because paprika is a colouring product, it can be adulterated with dyes, such as Sudan I, a carcinogenic agent. Microbiological safety refers to the absence of pathogens. Paprika can be contaminated with bacteria such as *Salmonella* spp., *Bacillus cereus* and *Escherichia coli*. The correct cleaning and drying methods can avoid the development of mycotoxins.[17]

Other adulterants that can be present in paprika products are other vegetables, mainly to enhance taste, weight and colour. Although adulteration with vegetables is less dangerous to human health than dyes, they are still considered as adulterants when the producer does not inform consumers of their presence on the label. Vegetable adulterants can be urucum (*Bixa orellana* L.), which is another colouring species, starch (*Zea mays* L.), black pepper (*P. nigrum* L.), ginger (*Zingiber officinale* Roscoe) and turmeric (*Curcuma longa* L.).[18,19] In Brazil, a health surveillance study with commercial samples[18] demonstrated residues of insects, rats and mites as contaminants to be monitored because as such residues indicate possible failures in the production chain.

11.3.3 Ornamental Plants

Capsicum species have a specific appearance with white to purple flowers and colourful fruits. Peppers were, and still are, appreciated for their beauty. Until the 1960s, chilli pepper seedlings and bushes were a popular Christmas gift due to their red bell-shaped fruits resembling a decorated Christmas tree. Ornamental cultivars of *Capsicum* are easy to cultivate indoors because they are small (up to 35 cm average height), relatively resistance to heat and drought,

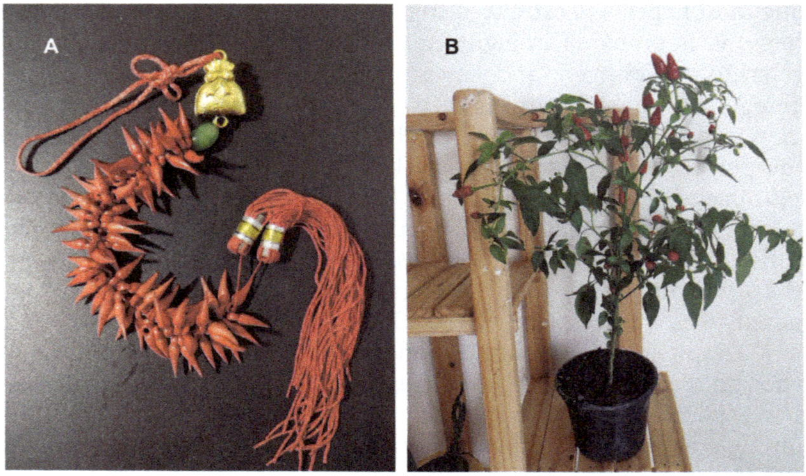

Figure 11.6 Peppers as (A) decorative jewellery and (B) a "Christmas tree" house plant. Image credit: Carlos Magno da Silva Antonio.

have a short cropping time and give fruits more than once a year. Their use as ornamental plants is not only attributed to their appearance, but also to other cultural features that consider chilli peppers as an amulet capable of absorbing bad energy, withdrawing bad spirits, preventing hurtful gossip from envious tongues and bringing good fortune. Even today, jewels and other decorative objects are made in the form of peppers (Figure 11.6).[20,21]

Capsicum species are diverse, both morphologically and genetically. Similar to the several choices regarding pungency, colour and aroma in industrial applications of peppers, there are also many ornamental options. For ornamental purposes, *Capsicum* can be divided into tree categories.[20]

- Potted plants: easy to propagate from seeds (compatible with self-pollination and cross-pollination); short cropping times; heat and drought tolerance (suitable for indoor cultivation); a polychotomous habitat; vibrant fruit and flower colours; 10–15 cm height.
- Border and garden plants: heat and drought tolerant; vibrant colours during summer; erect fruits above the foliage; can be used in culinary applications; polychotomous habitat with basal branches for use in garden borders (Figure 11.7); smooth surface fruits; erect flower pedicels.
- Cut stems: used for flower arrangements: long stem with clustered, pointed and erect fruits (Figure 11.7); usually six fruits per cluster, with smooth and shiny surfaces; popularly used to make ristras, a long arrangement of fresh or dry peppers.

All these ornamental uses are important alternative sources of income in the local bioeconomy of less developed countries where these species naturally grow.

Figure 11.7 Basal-branched (P138B) and fruit-clustered (P119) cultivars of *Capsicum*. Reproduced from ref. 22, https://doi.org/10.1590/S0102-053620160000100015, under the terms of the CC BY 4.0 license, https://creativecommons.org/licenses/by/4.0/. Accessions from *Capsicum* Active Germplasm Bank of Embrapa Temperate Agriculture.

11.3.4 Weight Control

In addition their nutritional value, *Capsicum* peppers can also be used in preparations to control obesity. Being overweight can lead to severe clinical conditions, such as diabetes, cardiovascular disease and osteoarthritis. Although obesity is a treatable and preventable disease (in most cases it can be avoided or reversed by adopting a healthier lifestyle), it was estimated in 2016 that nearly 13% of world's adult population is obese. Obesity can be treated by regular exercise, a specific diet rich in polyunsaturated fatty acids and low in carbohydrates (a ketogenic diet) or the use of drugs to reduce food intake following medical guidance. However, pharmacotherapy with synthetic drugs can induce liver diseases, hypertension and worsen coronary syndromes.

Phytotherapy has fewer side-effects when used to aid weight loss in people with obesity. The capsaicinoids and other phenolic compounds (*e.g.* catechins) found in *Capsicum* fruits inhibit lipogenesis, promote lipolysis (the degradation of fat), reduce adipogenesis and help to burn energy (thermogenesis). Formulations of phytotherapeutic medicines for long-term weight loss and management include Capsifen® and Capsimax®, which should not be used without medical guidance (Figure 11.8). These medicines maintain a capsaicinoid concentration of 4 mg per day, a physiologically relevant dose, but avoid the adverse symptoms related to pungency, such as gastric inflammation and respiratory allergies. Treatment with this concentration of capsaicinoids for 28 days has been proved to cause a significant reduction in body weight, body mass index and waist to hip ratio, which are indicators of overweight.[23]

The thermogenesis action of capsaicinoids occurs *via* the activation of TRPV1. This receptor is responsible for the burning and heat sensations of

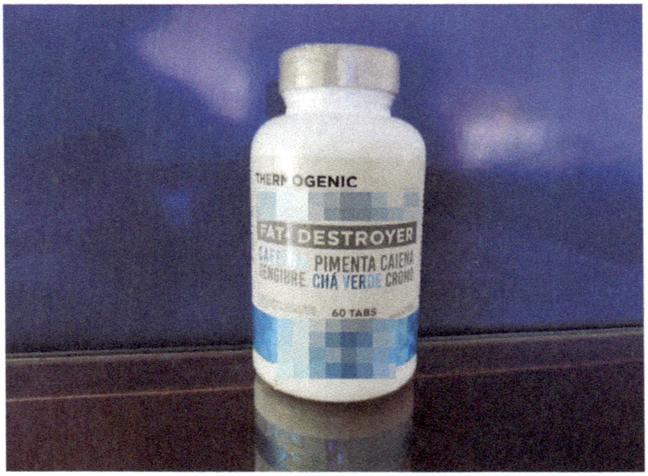

Figure 11.8 Commercial thermogenic product for weight control with *Capsicum annuum* cayenne extracts in its formula. Image credit: Carlos Magno da Silva Antonio.

chilli peppers. As one of the nerve receptors involved in thermal control in the human body, the activation of TRPV1 induces thermogenesis in brown adipose tissue. This type of tissue is used by the body to provide energy at low temperatures, but its excessive accumulation is one of the key factors in the development of obesity. TRPV1 is constantly activated by capsaicinoids during treatment, inducing thermogenesis in brown adipose tissue and resulting in weight loss.[24,25] The supplementation of capsaicinoid-based drugs alone as an agent of thermogenesis does not work and it is mandatory to maintain other measures, such as regular exercise.

The effect of capsaicinoids on thermogenesis is also used to improve physical performance. Because the activation of TRPV1 burns adipose tissue, more energy is available to the muscles. In addition, as TRPV1 (a cation channel) is activated, more Ca^{2+} is released in the sarcoplasmic reticulum of muscle cells, reducing fatigue. However, the ergogenicity of capsaicinoids in humans is not known precisely and different studies report different improvements in performance. Research indicates that the efficacy of these formulations depends on factors such as the type of consumption (acute or prolonged), diet and exercise duration.[26,27]

11.3.5 Cosmetics

There are at least 19 different cosmetic products that use *Capsicum*-derived ingredients in their formulations. Most of the *Capsicum*-derived products are obtained from *C. annuum* and *Capsicum frutescens* as whole fresh fruit, extracts, oleoresins and dried powders. The *Capsicum*-derived ingredients in cosmetics can be used to control seborrheic dermatitis and psoriasis, and as antifungal or antimicrobial agents, antioxidants, astringents, rubefacients, fragrances, flavourings, dyes or analgesics.

Within these preparations, the goal of the application of *Capsicum*-derived ingredients is to access their bioactive chemical classes, such as the capsaicinoids, carotenoids, phenolic compounds and vitamins. For instance, isopropanol pepper extracts – which are rich in cinnamic, coumaric, ferulic and caffeic acids – can be used as preservative agents in cosmetics because they have antibacterial activity against pathogens such as *Listeria monocytogenes*, *Bacillus cereus*, *Staphylococcus aureus* and *Salmonella typhimurium*. Other sweet pepper extracts are also rich in polyphenols and carotenoids with activity against *B. cereus* and *E. coli*.[28]

The capsaicin concentration in *Capsicum*-derived products used in cosmetics ranges from 0.0003 to 5% w/w. In addition, the total content of capsaicinoids must not exceed 1 g per 100 g of the total wet weight of the cosmetic due to the high risk of intoxication. It is not common to find beauty cosmetics with capsaicinoids, although *Capsicum*-derived products are not banned from formulations. One of the major concerns is that capsaicin-containing products should not have contact with the eyes and mucosal tissues. Most of the capsaicin-containing formulations are hair products, body/skin lotions or soaps (Figure 11.9).[29] The use of capsaicin in such

Figure 11.9 Shampoo product with *Capsicum annuum* extract in its formulation. Image credit: Carlos Magno da Silva Antonio.

products is intended to promote blood circulation of the hair follicles to treat alopecia.

A curious application of capsaicin in beauty cosmetics is its use in lipsticks. These products state that the presence of chilli pepper extracts will pump up the lips *via* micro-stimulation of the blood vessels. However, none of the safety control agencies corroborate this effect. In addition, the pumping up effect could be due to irritation of the mucosal tissues by capsaicin.

11.4 Concluding Remarks

The *Capsicum* genus is fascinating botanically. The chemical composition of *Capsicum* species is unique and has been of interest to human civilisations from ancient times to the modern day, with *Capsicum* peppers applied for many purposes. Their chemical profile makes them a remarkable source of essential nutrients, but also a tasty ingredient and sometimes an ornament. The applications of this crop are vast, yet, even with the efforts of the scientific community to fully uncover the potential of this genus, there is much to be achieved. Quality control and standardisation procedures must be developed to better explore *Capsicum*-based products, even for the well-known spice paprika. Cosmetic and therapeutic pepper-based products are underrated as a result of the pungency of several species and varieties.

References

1. Observatory of Economic Complexity, Peppers (Capsicum, Pimenta) fresh or Chill, 2021, https://oec.world/en/profile/hs92/peppers-capsicum-pimenta-fresh-or-chilled?growthSelector=value2 (accessed August 2021).
2. Crop Information: Peppers, http://www.fao.org/land-water/databases-and-software/crop-information/pepper/en/ (accessed August 2021).
3. FAOSTATS, http://www.fao.org/faostat/en/#data/QC (accessed August 2021).
4. M. F. Lima, S. I. C. Carvalho, C. F. Ragassi, L. B. Bianchetti, F. G. Faleiro and F. J. B. Reifschneider, *Genet. Mol. Res.*, 2017, **16**, 1–18.
5. S. M. Khasanah, M. Maksum and E. Suwondo, *agriTECH*, 2020, **40**, 57.
6. G. T. Sigurdson, P. Tang and M. M. Giusti, *Annu. Rev. Food Sci. Technol.*, 2017, **8**, 261–280.
7. M. F. Mohamad, D. J. Dailin, S. E. Gomaa, M. Nurjayadi and H. El Enshasy, *Int. J. Life Sci. Biotechnol. Pharma Res.*, 2019, **8**, 3161–3166.
8. T. G. Berke and S. C. Shieh, in *Handbook of Herbs and Spices*, ed. K. V. Peter, Woodhead Publishing Limited, Philadelphia, 2nd edn, 2012, ch. 7, pp. 116–130.
9. *Handbook of Eating and Drinking*, ed. H. L. Meiselman, Springer, Rockport, 2020.
10. Z. Shi, M. Riley, A. W. Taylor and A. Page, *Int. J. Obes.*, 2017, **41**, 1074–1079.

11. M. Bonaccio, A. Di Castelnuovo, S. Costanzo, E. Ruggiero, A. De Curtis, M. Persichillo, C. Tabolacci, F. Facchiano, C. Cerletti, M. B. Donati, G. de Gaetano, L. Iacoviello, A. Bonanni, F. Gianfagna, T. Di Prospero, J. Vermylen, I. De Paula Carrasco, S. Costanzo, A. Spagnuolo, D. Assanelli, V. Centritto, M. Olivieri, A. Gialluisi, F. Gianfagna, S. Magnacca, B. Izzi, A. Marotta, F. Noro, F. De Lucia, F. Bracone, S. Dudiez, L. Rago and T. Panzera, *J. Am. Coll. Cardiol.*, 2019, **74**, 3139–3149.
12. Z. Cao, L. Zhou, J. Bi, J. Yi, Q. Chen, X. Wu, J. Zheng and S. Li, *J. Sci. Food Agric.*, 2016, **96**, 3596–3603.
13. A. Topuz, C. Dincer, K. S. Özdemir, H. Feng and M. Kushad, *Food Chem.*, 2011, **129**, 860–865.
14. F. Karatas, A. Baysar and M. Alpaslan, *J. Food Process. Preserv.*, 2017, **41**, 1–5.
15. H. G. Daood, G. Palotás, G. Palotás, G. Somogyi, Z. Pék and L. Helyes, *Food Res. Int.*, 2014, **65**, 231–237.
16. V. Szucs and E. Szabo, *J. Hyg. Eng. Des.*, 2007, **521**, 5–9.
17. E. D. van Asselt, J. L. Banach and H. J. van der Fels-Klerx, *Food Control*, 2018, **83**, 7–17.
18. L. F. de Pauli-Yamada, C. I. Aquino, A. M. Silva, M. A. M. Marciano, E. C. de Mattos and M. D. Nogueira, *Vigilância Sanitária em Debate Soc. Ciência Tecnol.*, 2021, **9**, 123–128.
19. M. M. Oliveira, J. P. Cruz-Tirado, J. V. Roque, R. F. Teófilo and D. F. Barbin, *J. Food Compos. Anal.*, 2020, **87**, 103403.
20. J. R. Stommel and P. W. Bosland, in *Flower Breeding and Genetics: Issues, Challenges and Opportunities for the 21st Century*, ed. N. O. Anderson, Springer, Netherlands, 2006, ch. 21, pp. 561–599.
21. G. D. N. Costa, B. M. P. Da Silva, Â. C. D. A. Lopes, L. C. B. Carvalho and R. L. F. Gomes, *Rev. Caatinga*, 2019, **32**, 566–574.
22. R. S. Neitzke, S. Z. Fischer, C. S. Vasconcelos, R. L. Barbieri and R. O. Treptow, *Hortic. Bras.*, 2016, **34**, 102–109.
23. A. Joseph, F. John, J. V. Thomas, S. Das, Sivadasan, B. Maliakel, R. Mohan and I. M. Krishnakumar, *J. Diet. Suppl.*, 2020, **0**, 1–19.
24. C. S. Padilha, F. Billaut, C. Figueiredo, V. L. G. Panissa, F. E. Rossi and F. S. Lira, *Int. J. Sports Med.*, 2020, **41**, 1061–1066.
25. H. H. Jang, J. Lee, S. H. Lee and Y. M. Lee, *Sci. Rep.*, 2020, **10**, 1–12.
26. V. E. L. de Moura e Silva, J. M. Cholewa, F. Billaut, R. Jäger, M. C. de Freitas, F. S. Lira and F. E. Rossi, *Int. J. Sports Physiol. Perform.*, 2020, 1–10.
27. A. E. von Ah Morano, C. S. Padilha, V. A. M. Soares, F. A. Machado, P. Hofmann, F. E. Rossi and F. S. Lira, *Nutrients*, 2021, **13**, 1–10.
28. N. Baenas, M. Belović, N. Ilic, D. A. Moreno and C. García-Viguera, *Food Chem.*, 2019, **274**, 872–885.
29. L. F. Scacheti, N. C. Matos, L. Mallafati and F. F. Navarro, *Rev. Cienc. Farm. Basica Apl.*, 2011, **32**, 369–374.

Subject Index